BASIC

ELECTRICITY/ELECTRONICS

VOLUME 5

Motors & Generators— How They Work

By
Training & Retraining, Inc.

Second Edition Revised by
Joseph A. Risse, P.E.

Howard W. Sams & Co., Inc.
4300 WEST 62ND ST. INDIANAPOLIS, INDIANA 46268 USA

Edited by: *Frank N. Speights*
Illustrated by: *R. E. Lund*

Printed in the United States of America.

Acknowledgments

Grateful acknowledgment is made to all those who participated in the preparation, compilation, and editing of this series. Without their valuable contributions, this series would not have been possible.

In this regard, prime consideration is due Bernard C. Monnes, Educational Specialist, Navy Electronics School, for his excellent contributions in the areas of writing, editorial organization, and final review of the entire series. The finalization of these volumes, both as to technical content and educational value, is due principally to his tireless and conscientious efforts.

Grateful appreciation is also extended to Lt. Loren Worley, USN, and Ashley G. Skidmore, BUSHIPS, Dept. of the Navy, for their original preparatory contributions and coediting of this series. We also want to thank Irene and Don Koosis, Raymond Mungiu, George V. Novotny, and Robert J. Brite for their technical writing and contribution to the programmed method of presentation. Special thanks to Robert L. Snyder for his initial preparation and organizational work on the complete series.

Credit for the initial concept of this programmed learning series goes to Stanley B. Schiffman, staff member of Training & Retraining, Inc.

Finally, special thanks are due the Publisher's editorial staff for invaluable assistance beyond the normal publisher-author relationship.

TRAINING & RETRAINING, INC.

Contents

CHAPTER 3

CHAPTER 4

CHAPTER 5

CHAPTER 6

CHAPTER 7

CHAPTER 8

Introduction

This volume, the fifth and last in the series, is concerned with the principles of ac and dc motors and generators. The text is designed to provide the reader with a sound understanding of the fundamentals of motors and generators. The characteristics of each kind of machine are presented so that the student can relate each kind to actual applications. In this way, reader interest is maintained, and the information presented is of practical value.

WHAT YOU WILL LEARN

You will be shown how the operation of motors and generators depends on the basic electrical principles you have already learned. You will see how these principles are put to work to perform specific tasks with electrical machines.

First you will be introduced to the ways in which mechanical energy can be transformed into electrical energy, and vice versa. Such terms as commutator, slip rings, torque, left-hand rule, brush, armature, and others are explained.

Dc generators are discussed in detail. Construction details are given, including bearings, armature cores, etc. Wave windings as well as simplex, duplex, and triplex lap windings are explained. The characteristics of separately excited, shunt, series, and compound generators are described, along with losses (copper loss, eddy-current loss, and hysteresis loss). You will learn about armature reaction and the methods used to counteract it. Methods for paralleling dc generators are given, and maintenance techniques are described.

You will learn that dc motors are similar in construction to dc generators. Counter emf and armature reaction are discussed, as are characteristics of shunt, series, and compound motors. Some various kinds of dc starters and controllers that are described include starter boxes, faceplate starters, and drum controllers. Also included in the discussions are electronic speed controls, brushless motors, and different kinds of interference from dc motors and how to reduce this problem. Some common methods of motor speed control are listed.

The text on alternators (ac generators) explains synchronous, induction, single-phase, and polyphase types. Armature reaction frequency control and voltage regulation are also discussed and methods of connecting several alternators in parallel are described.

The application of three-phase electromagnetic fields to motor operation is explained. You will learn about polyphase synchronous and induction motors. You will also learn about several kinds of single-phase motors, including shaded-pole, split-phase, capacitor, repulsion, repulsion-induction, and universal motors. An energy-saver motor control is also described.

The generation and distribution of electrical power by three-phase systems along with the relationships of voltage, current, and power in both wye- and delta-connected systems are described.

You will learn about electrical and electronic devices that are used to convert electric power from one form to another (ac to dc, dc to dc of a different voltage, etc.). Finally, you will learn about servo systems including electron-tube and solid-state servo amplifiers, open- and closed-servo systems, servo motors, ac and dc servo amplifiers, and synchromechanisms.

WHAT YOU SHOULD KNOW BEFORE YOU START

Before you start to study this book, it is essential that you have a good background in the principles of electricity and electronics, including the fundamentals of tube and transistor circuits and test equipment. This background can be obtained by studying the first four volumes of this series. With the proper background, you should have no trouble understanding this text. All new terms are carefully defined. Enough math is used to give a precise interpretation of important principles, and if you know how to add, subtract, multiply, and divide, the mathematical expressions will give you no trouble.

WHY THE TEXT FORMAT WAS CHOSEN

During the past few years, new concepts of learning have been developed under the common heading of programmed instruction. Although there are arguments for and against each of the several formats or styles of programmed textbooks, the value of programmed instruction itself has been proved to be sound. Most educators now seem to agree that a style of programming should be developed to fit the needs for teaching each particular subject. To help you progress successfully through this volume, a brief explanation of the programmed format follows.

Each chapter is divided into small bits of information and presented in a sequence that has been proved best for learning purposes. Some of the information bits are very short—a single sentence in some cases. Others may include several paragraphs. The length of each presentation is determined by the nature of the concept being explained and the knowledge the reader has gained up to that point.

The text is designed around two-page segments. Facing pages include information on one or more concepts, complete with illustrations designed to clarify the word descriptions used. Self-testing questions are included in most of these two-page segments. Many of these questions are in the form of statements that require you to fill in one or more missing words; other questions are either multiple-choice or simple essay types. Answers are given on the succeeding page, so you will have the opportunity to check the accuracy of your response and verify what you have or have not learned before proceeding. When you find that your answer to a question does not agree with that given, you should restudy the information to determine why your answer was incorrect. As you can see, this method of question–answer programming ensures that you will advance through the text as quickly as you are able to absorb what has been presented.

The beginning of each chapter features a preview of its contents and, then, a review of the important points is contained at the end of the chapter. The preview gives you an idea of the purpose of the chapter—what you can expect to learn. This helps to give practical meaning to the information as it is presented. The review at the end of the chapter summarizes its content so that you can locate and restudy those areas which have escaped your full comprehension. And, just as important, the review is a definite aid for retention and recall of what you have learned.

HOW YOU SHOULD STUDY THIS TEXT

Naturally, good study habits are important. You should set aside a specific time each day to study in an area where you can concentrate without being disturbed. Select a time when you are at your mental peak, a period when you feel most alert.

Here are a few pointers you will find helpful in getting the most out of this volume.

1. Read each sentence carefully and deliberately. There are no unnecessary words or phrases; each sentence presents or supports a thought that is important to your understanding of electricity and electronics.

2. When you are referred to or come to an illustration, stop at the end of the sentence you are reading and study the illustration. Make sure you have a mental picture of its general content. Then, continue reading, returning to the illustration each time a detailed examination is required. The drawings were especially planned to reinforce your understanding of the subject.

3. At the bottom of most right-hand pages, you will find one or more questions to be answered. Some of these contain "fill-in" blanks. Since more than one word might logically fill a given blank, the number of dashes indicates the number of letters in the desired word. In answering the questions, it is important that you actually do so in writing, either in the book or on a separate sheet of paper. The physical act of writing the answers provides a greater retention than merely thinking the answer. However, writ-

ing will not become a chore since most of the required answers are short.

4. Answer all questions in a section before turning the page to check the accuracy of your responses. Refer to any of the material you have read if you need help. If you don't know the answer even after a quick review of the related text, finish answering any remaining questions. If the answers to any questions that you skipped still haven't come to you, turn the page and check the answer section.

5. When you have answered a question incorrectly, return to the appropriate paragraph or page and restudy the material. Knowing the correct answer to a question is less important than understanding **why** it is correct. Each section of new material is based on previously presented information. If there is a weak link in this chain, the later material will be more difficult to understand.

6. In some instances, the text describes certain principles in terms of the results of simple experiments. The information is presented so that you will gain knowledge whether you perform the experiments or not. However, you will gain a greater understanding of the subject if you do perform the suggested experiments.

7. Carefully study the review, "What You Have Learned," at the end of each chapter. This review will help you gauge your knowledge of the information in the chapter and will actually reinforce your knowledge. When you run across statements that you don't completely understand, reread the sections relating to these statements, and recheck the questions and answers before going to the next chapter.

This volume has been carefully planned to make the learning process as easy as possible. Naturally, a certain amount of effort on your part is required if you are to obtain the maximum benefit from the book. However, if you follow the pointers just given, your efforts will be well rewarded, and you will find that your study of electricity and electronics will be a pleasant and interesting experience.

1

Understanding Basic Principles

what you will learn

In this chapter, you will learn about the basic principles of motors and generators. You will learn how generators convert mechanical energy into electrical energy. You will find that generators can provide an ac or dc output depending on whether the generated current is taken from the generator through slip rings or a commutator. You will also learn how ac and dc motors convert electrical energy into mechanical energy.

SOURCES OF ELECTRICITY

Earlier you learned that electrical energy is usually supplied by batteries or electrical power plants. Batteries are generally used where portability is desired and small amounts of current are needed. However, the voltage and current that can be developed by one cell are small. To increase the voltage, a number of cells must be connected in series. To increase the current capacity, either the cell must be made larger or a number of cells must be connected in parallel. A battery that supplies both high current and high voltage is bulky and expensive.

A voltage drop occurs when dc is transmitted over long distances, so batteries must be located near the place where the electrical energy is used. Batteries are usually used when a portable source of small values of dc is needed.

PRODUCTION OF ELECTRICAL ENERGY

Large amounts of electrical energy are usually supplied by **generators** in power plants. A generator is defined as a **machine that converts mechanical energy into electrical energy.**

Mechanical energy is converted into electrical energy by **induction.** The example in Fig. 1-1 shows how a voltage is generated by induction. Your hand supplies the mechanical energy needed to move the coil, and the multimeter detects the electrical energy produced.

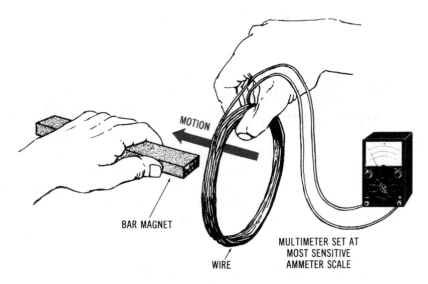

BAR MAGNET

MOTION

WIRE

MULTIMETER SET AT
MOST SENSITIVE
AMMETER SCALE

Fig. 1-1. Generating a voltage by induction.

If an electric conductor is moved through a magnetic field in such a way that it cuts the lines of force, a voltage is generated, or induced, in the conductor. The induced voltage is greatest when the conductor moves at right angles to the magnetic field, and is zero when the conductor moves parallel to the lines of force.

If the moving conductor is connected to a complete electric circuit, an electric current will flow in the conductor and in the circuit. This means that the **mechanical energy** used in moving the conductor through the magnetic field is converted into **electrical energy** which moves the current through the circuit.

Factors Determining Voltage

The amount of voltage generated in a moving conductor is determined by: (1) **The speed at which the conductor passes through the magnetic field.** Greater speed causes the conductor to cut more lines of force per second. (2) **The strength of the magnetic field.** A stronger field provides more magnetic lines of force. (3) **The number of loops of wire.** Each additional loop of wire increases the number of conductors in which a voltage may be induced. Since the loops are in series, the generated voltages in each are additive.

Conditions for Generating a Voltage

Voltage is generated by induction when a conductor is moved through a magnetic field, or when a magnetic field moves in relation to a stationary conductor located in the field. In other words, **voltage is generated when a conductor and a magnetic field move relative to each other.**

In a generator, the conductor can move, the field can move, or both can move. All of these possibilities, as used in practical generators, are discussed in this volume.

The amount of voltage induced in a conductor by a magnetic field depends, among other factors, on the distance between the magnetic pole and the conductor. When the distance between the magnetic pole and the conductor decreases, the magnetic field through which the conductor is moving is stronger. The conductor then cuts through more magnetic lines of force per given distance of movement, and a higher voltage is produced.

Q1-1. A generator converts mechanical energy into electrical energy by the principle of _ _ _ _ _ _ _ _ _ .

Q1-2. The amount of voltage generated by induction increases as the _ _ _ _ _ between the conductor and the magnetic pole increases.

Q1-3. Name three ways to generate an emf by induction.

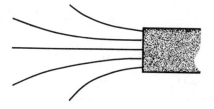

Fig. 1-2. Field strength is greatest near a pole.

Transmission of Mechanical Energy to the Generator

Mechanical energy is usually transmitted to a generator by a shaft. When a coil, called the **armature,** is mounted on the shaft and the assembly is rotated in a magnetic field, an **emf** (voltage) is induced in the armature.

Energy from waterfalls, wind, tides, or high-pressure steam can be used to turn a **turbine** that will rotate the shaft. A turbine is a machine that converts water or steam pressure into the rotation of a shaft (Fig. 1-3).

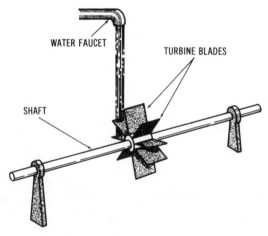

WATER FAUCET

TURBINE BLADES

SHAFT

Fig. 1-3. A simple water turbine.

POWER PLANT LOCATION

The location of power plants depends on the availability of a plentiful source of energy that can be used to rotate the generator shaft.

In steam power plants, high-pressure steam is directed against the blades of a turbine which rotates the generator shaft. Heat from coal, nuclear energy, or oil is used to boil water and create high-pressure steam. This kind of power plant requires large amounts of water, as well as fuel, in order to produce steam.

Hydroelectric power generating plants are located at the base of waterfalls or dams. Water flowing over the waterfall or released by the dam is directed against the turbine blades, causing the shaft of the generator to rotate.

WHAT IS A MOTOR?

Electrical energy produced by generators is used in many ways. One of the most important uses is to turn motors that operate machinery and appliances. A **motor** is a machine that converts **electrical energy** into **mechanical energy.** This definition is the inverse of the definition of a generator.

When electrical energy is supplied to a motor, current flows through the armature. The current creates a magnetic field surrounding the armature. This field interacts with a stationary magnetic field. The interaction of the two magnetic fields creates a twisting force, or **torque,** that causes the shaft of the motor to rotate.

Q1-4. Name three methods that are used to turn a generator shaft.

Q1-5. When water or high-pressure steam is directed against the turbine blades, the generator shaft will _____ .

Q1-6. What will happen if the force with which the water strikes the turbine blades is increased?

Q1-7. A twisting force is called _____ .

Q1-8. A motor is a machine that converts _____ _____ into _____ _____ .

Q1-9. A water wheel is a simple example of a _____ .

Converting Electrical Energy Into Mechanical Energy

The experiment shown in Fig. 1-4 demonstrates how forces are created in a motor. If you try the experiment, be careful not to close the switch until you are ready to observe the results. When you close the switch, you should do so for only a few seconds. This is because the battery is short-circuited when the switch is closed. If the switch is not reopened quickly,

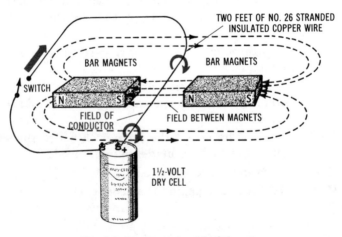

Fig. 1-4. Simple motor principle.

the battery will be drained of its electrical energy. When the switch is closed, the conductor should jump. When the switch is opened, the conductor should return to its original position.

The conductor jumps from between the magnets when the switch is closed because a magnetic field is created by the flow of current through the conductor. The magnetic field around the conductor causes the main magnetic field to be strengthened on one side of the conductor and weakened on the other. This creates a force which pushes the wire away from the stronger part of the field (Fig. 1-5).

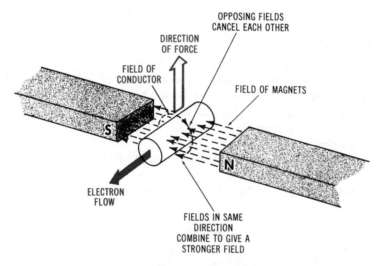

Fig. 1-5. Conductor in a magnetic field.

The direction of the magnetic field surrounding the conductor can be determined by applying the **left-hand rule** for straight conductors. Wrap the fingers of your left hand around the conductor so that your thumb points in the direction of electron flow through the conductor. Your fingers then show the direction of the magnetic field surrounding the conductor. External magnetic fields are considered to flow from the north pole to the south pole.

Q1-10. State the left-hand rule for straight conductors.

Q1-11. The difference in the strength of the magnetic field above and below the conductor results in a ──── .

What Is Torque?

Torque (pronounced "tork") is a turning force. In a motor, the conductor is formed into a coil and placed on a shaft that is free to rotate. When current flows through the coil, a magnetic field is produced. This magnetic field around the coil reacts with the stationary magnetic field, developing a torque and causing the shaft to turn.

Torque is calculated by multiplying the force times the distance from the center of rotation. The illustration given in Fig. 1-6 shows a two-foot ruler balanced at its center. When a

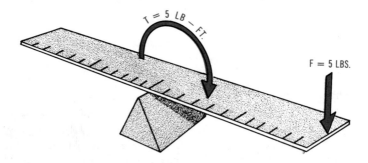

Fig. 1-6. Calculation of torque.

force of five pounds is applied to the right end of the ruler, the resulting torque is calculated as follows:

Torque = Force × Distance from center of rotation
    = 5 pounds × 1 foot
    = 5 pound-feet

Shown in Fig. 1-7 is a coil that is free to rotate in a magnetic field. The flow of current out of the left side of the coil causes

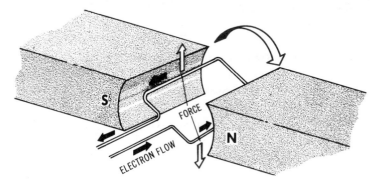

Fig. 1-7. Production of torque.

the magnetic field below the conductor to be strengthened and the field above to be weakened. This results in an upward force on the left side of the coil and causes a clockwise direction of rotation.

The flow of current into the right side of the coil causes the magnetic field below the conductor to be weakened and it causes the field above the conductor to be strengthened. This results in a downward force on the right side of the coil and a clockwise rotation.

The amount of torque generated depends on the strength of the two magnetic fields and on the distance of the sides of the coil from the center of rotation. The strength of the magnetic field around the coil will depend on the number of turns in the coil, the current through the coil, the core material of the coil, and so forth.

Q1-12. If the ruler shown in Fig. 1-6 were six feet long and had a force of ten pounds applied to the right end, how much torque would be developed?

Q1-13. What factors determine the amount of torque on the coil shown in Fig. 1-7?

Q1-14. What factors determine the strength of the magnetic field around the coil shown in Fig. 1-7?

Q1-15. Torque is calculated by multiplying the _ _ _ _ _ times the total _ _ _ _ _ _ _ _ from the center of rotation.

AC AND DC GENERATORS

The output from the armature coil of any generator is an ac voltage. As the coil shown in Fig. 1-8 rotates at a constant speed, it cuts more or fewer magnetic lines of force per second,

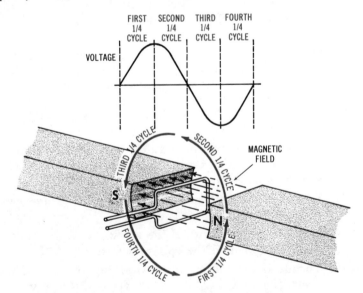

Fig. 1-8. Generation of a sine-wave voltage.

depending on its position at any particular instant. When the coil is moving at right angles to the magnetic field, it is cutting a maximum number of lines of force per second. Therefore, the voltage induced in the coil increases until, when the coil is moving at right angles to the field, the voltage is maximum. Then, as the coil continues to rotate, it cuts fewer and fewer lines of force per second until it is moving parallel to the magnetic field. At that point, no voltage is induced. In the third quarter cycle, the conductor cuts the lines of force in the opposite direction, so the induced voltage has the opposite polarity and again rises to a maximum. In the fourth quarter cycle, the voltage again decreases to zero. This is shown in Fig. 1-8.

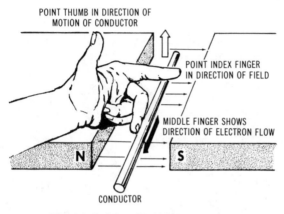

Fig. 1-9. Left-hand rule for generators.

The **left-hand rule** for generators is shown in Fig. 1-9. This is an easy way of remembering the relationship between the direction of the magnetic field, the direction of motion of the conductor, and the direction of the induced current.

In a generator, the shaft rotation determines the direction that the conductor moves. The direction of the magnetic lines of force is from the north pole to the south pole. If the polarity of a magnet is unknown, it can be determined by using a compass. The south end of the compass needle will point to the north pole of the magnet.

Q1-16. A(an) _ _ voltage is induced in a rotating generator coil.

Q1-17. Describe the left-hand rule for generators.

The Ac Generator

The output from the armature coils of any generator is alternating current. In order to take the ac output of the generator

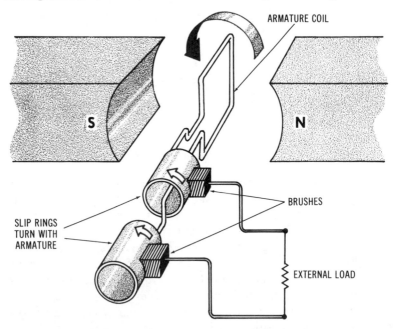

Fig. 1-10. A simple ac generator.

from the armature coils, the ends of the coils are connected to **slip rings** which rotate with the armature. Stationary conductors called **brushes** provide a sliding contact on the rotating slip rings. In this way, the brushes and slip rings provide a connection between the armature coils in the generator and any external load that is being furnished power by the generator.

The Dc Generator

In a dc generator, the ac output of the armature coils is converted to pulsating direct current by the use of a **commutator** in the place of slip rings. The output of a basic dc generator is shown in Fig. 1-11. This pulsating dc output is obtained

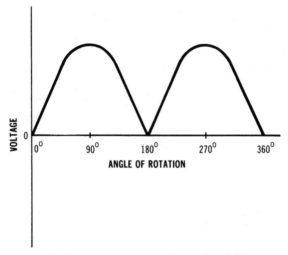

Fig. 1-11. Output of a basic dc generator.

because the contacts reverse the connections to the armature coil every half cycle when the voltage is zero and is about to change polarity.

Q1-18. In an ac generator, _ _ _ _ _ _ _ _ _ are used to take the output of the generator from the armature coils.

Q1-19. _ _ _ _ _ _ _ are used to make a sliding contact on the rotating slip rings.

Q1-20. In a dc generator, a _ _ _ _ _ _ _ _ _ _ is used to convert the ac output of the armature coils into pulsating dc.

What Is a Commutator?

A basic commutator is simply a slip ring split into two semi-circular halves, called **segments.** The segments are insulated from each other and from the shaft. One end of the armature coil is connected to one segment and the other end of the coil

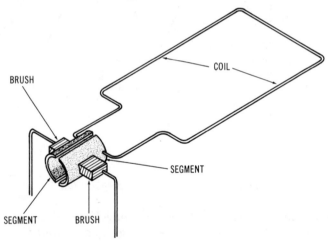

Fig. 1-12. A basic commutator.

is connected to the other segment. Two brushes touch opposite sides of the commutator. As the commutator turns, the two sides of the coil are short-circuited for a moment as the brushes touch both segments of the commutator at once (Fig. 1-13). Then, the connections are reversed.

If the armature coil is short-circuited by the brushes while an emf is being induced in the coil, a heavy current will flow in the armature coil. This is because the circuit formed by the coil and the brushes has a very low resistance. This excessive current may cause serious damage to the armature coils. Also,

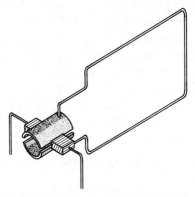

Fig. 1-13. Coil short-circuited by brushes.

if the commutator of a dc generator is not adjusted to reverse the armature coil connections at the moment when the induced emf is zero, the output of the generator will not be direct current. Instead, it will be partially alternating current, as shown in Fig. 1-14.

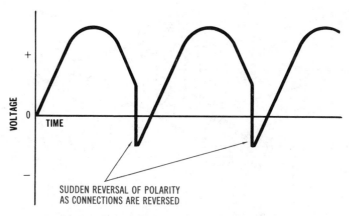

Fig. 1-14. Generator output when switching occurs at the wrong time.

Q1-21. At what point in the ac cycle should the commutator reverse the coil connections?

Q1-22. Give two reasons why proper adjustment of a commutator is important.

DC MOTORS

Dc motors also have commutators. In fact, a basic dc generator can also act as a dc motor. When a dc current passes
through the coil, the current creates a magnetic field. The north
pole of the coil is attracted to the south pole of the outside magnetic field, and the south pole of the coil is attracted to the north
pole of the outside field. Thus, the coil rotates.

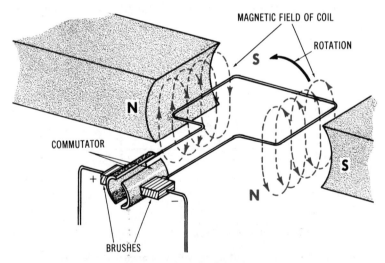

Fig. 1-15. A basic dc motor.

You can analyze the magnetic forces acting on the simple
coil by using the left-hand rule for conductors. Fig. 1-15 shows
the field around the coil.

When the coil reaches the position shown in Fig. 1-16, the brushes touch both commutator segments at the same time. No current flows in the coil and there is no turning force on the coil. However, the coil is turning and its inertia, which tends

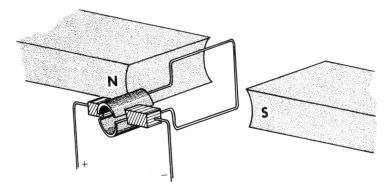

Fig. 1-16. Motor coil at dead center.

to keep it turning, carries it past dead center. As the coil rotates past dead center, the commutator reverses the direction of current flow through the coil. The polarity of the magnetic field around the coil is reversed and the pole of the coil that is next to the south pole of the external field now becomes the south pole of the coil. The new south pole of the coil is now attracted toward the north pole of the external field, so the coil keeps on rotating. The same switching process is repeated again when the coil has rotated another 180°.

In Fig. 1-16, the commutator causes a momentary short circuit across the power source when the coil is at dead center. You can see why this simple commutator-brush arrangement is not used in practical motors. The basic principle of all commutators is the same, however. As you have seen, the commutator in a dc motor converts the dc power supplied to the motor into ac for the armature coil.

Q1-23. Would the motor just described rotate if you supplied ac to the coil through slip rings?

Q1-24. The commutator in a dc motor changes direct current into _ _ _ _ _ _ _ _ _ _ _ _ _ _ _ _ _ _.

Q1-25. There is no magnetic field developed around the coil when it is at _ _ _ _ _ _ _ _ _ _.

WHAT YOU HAVE LEARNED

1. A generator is a machine that converts mechanical energy into electrical energy by induction.
2. For induction to occur, there must either be a conductor moving in a stationary magnetic field, a moving magnetic field surrounding a stationary conductor, or a moving magnetic field containing a moving conductor.
3. The strength of the induced emf depends on how fast magnetic lines are being cut by conductors, the strength of the magnetic field, the number of conductors in which an emf is being induced, and the distance between the source of the magnetic field and the conductor.
4. The direction of the induced emf depends on the direction that the conductor is moving and the direction of the field through which it is moving.
5. Alternating current is generated in the armature coils of all generators.
6. The output of a generator will be alternating current if the output of the rotating armature coils is removed through slip rings and brushes.
7. The output of a generator will be direct current if the output of the rotating armature coils is removed through a commutator and brushes.
8. Motors are machines that convert electrical energy into mechanical energy.
9. A dc motor is supplied with dc voltage and current which is converted into alternating current by the commutator for use in the armature coils.
10. An ac motor does not need a commutator.

2

DC Generators

what you will learn

In this chapter, you will learn about dc generator construction. You will learn to recognize each part and know its function. You will be able to recognize effects within the generator that waste power and how to minimize them. You will also learn the characteristics of the various kinds of dc generators.

CONSTRUCTION

The major parts of a dc generator are the frame, end bells, pole pieces, shaft, armature assembly, commutator assembly, and brush assembly. Fig. 2-1 shows the construction of a dc generator.

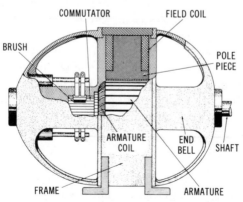

Fig. 2-1. Cutaway view of a dc generator.

Frame (Yoke)

The **frame**, or yoke, supports the generator. The frames of most modern generators are constructed of steel because this metal is an excellent conductor of magnetic lines of force. When the magnetic circuit is completed through a good magnetic conductor, as shown in Fig. 2-2, the field between the poles is stronger.

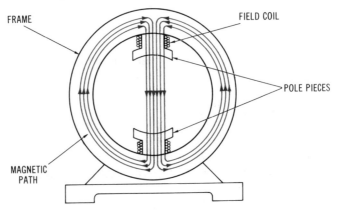

Fig. 2-2. Magnetic path in a generator frame.

At one time, most generator frames were made of cast iron. However, cast iron is heavier than steel and has a poorer **permeability** (ability to conduct magnetic lines of force), so it is used less frequently now. The frames of large generators are made of steel castings; those of smaller generators are made of rolled sheet steel. The frame of the generator also includes a base or mounting brackets.

There are three types of frames: open, semiclosed, and closed. An **open frame** has the ends open so that air can circulate freely through it. A **semiclosed frame** has a wire screen or a metal grille in its end bells to prevent foreign matter from entering the machine. A **closed frame** has solid end bells, and the machine is airtight.

Pole Pieces

The **pole pieces** (also called pole shoes) of a generator are always used in pairs and are bolted to the frame. They support the field windings that are used to produce a north pole

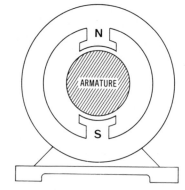

Fig. 2-3. Arrangement of two generator poles.

facing the armature on one side, with a south pole located directly opposite on the other side (Fig. 2-3).

More than one pair of poles are sometimes used in order to produce a stronger magnetic field. When this is done, the opposite poles are always next to each other. You can see by the illustration in Fig. 2-4 how this arrangement creates the strongest magnetic field through the armature.

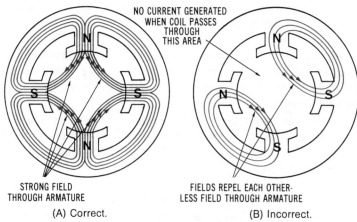

(A) Correct. (B) Incorrect.

Fig. 2-4. Arrangement of four generator poles.

Q2-1. The permeability of a material refers to how well the material conducts _.

Q2-2. Magnetic pole pieces are placed in a generator so that poles of (the same, opposite) polarity are next to each other.

End Bells and Bearings

The **end bells** are bolted to each end of the frame and contain the bearings that support the armature shaft. The three types of bearings most commonly used in generators are: **ring-oiled** (sleeve), **yarn-packed,** and **ball** bearings.

The ring-oiled, or sleeve, bearing contains an oil ring that rides on the shaft. As the shaft turns, oil is carried from a reservoir to the top of the shaft for distribution to the bearing surface.

The yarn-packed bearing consists of a bundle of wool yarn looped on the armature shaft. Both ends of the yarn loop are in the oil reservoir. Oil soaks into the yarn and lubricates the revolving shaft.

Both oil-ring and yarn-packed bearings are provided with oil seals at each end of the bearing in order to prevent oil from escaping and causing damage to the parts of the generator that carry electricity.

Ball bearings may be either the open or closed type and are packed with their own grease or lubricant. This type of bearing contains an outer ring, or race, and an inner ring. The outer ring does not move and is firmly held in position by the frame. The inner race rotates with the turning shaft. Between the inner and outer races are a number of very fine-machined steel balls that are packed in grease. This arrangement provides an almost friction-free rotation of the inner race while the outer race is stationary.

Armature and Commutator

As you learned in Chapter 1, the commutator used with a single coil produces a pulsating dc output. By using more coils and combining their outputs, a smoother waveform can be obtained.

When a second coil is added to the armature and is placed perpendicular to the first coil, the resulting output will be as shown in Fig. 2-5.

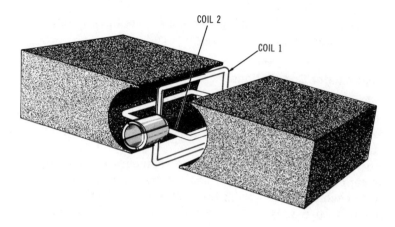

(A) Coil arrangement.

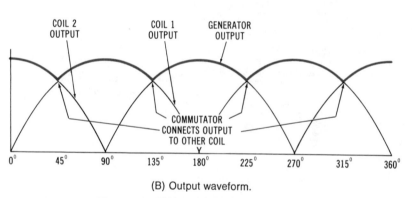

(B) Output waveform.

Fig. 2-5. Output of a two-coil dc generator.

Notice that an emf is induced at all times. Although the dc still pulsates, the output is smoother. In practical generators, many coils are added to the armature to produce a still smoother dc output.

Q2-3. Which type of bearing is usually lubricated with grease rather than oil?

Q2-4. A generator using two coils (does, does not) have a smoother output than one using one coil.

Commutator Connections

Each armature coil in a dc generator is connected to two commutator segments, with one segment for each end of the coil. A generator with many coils will also have many commutator segments.

In a practical commutator, the segments are usually made of copper. Mica is used to insulate the commutator segments from each other and from the commutator sleeve. The ends of the armature coils are connected to the raised portions of the

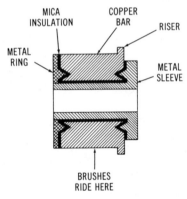

Fig. 2-6. Cross section of a commutator.

segments. The raised part of the segment is called a riser and is slotted to hold the coil ends. The brushes ride on the lower portions of the commutator segments.

Brush Assembly

The brush assembly shown in the illustration of Fig. 2-7 consists of the brush, brush holder, adjustable brush-holder springs, and pigtailed connections.

In normal operation, both the commutator segments and the brushes will experience wear. The adjustable brush-holder spring applies the brush pressure on the commutator and adjusts to compensate for this wear. However, the spring should apply only enough pressure to make a good connection between brush and commutator.

The pigtailed connections are braided copper wires that provide a low-resistance and flexible connection between the brush and the brush holder.

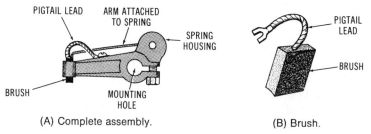

(A) Complete assembly.   (B) Brush.

Fig. 2-7. A brush assembly.

Brush Materials

The type and operating conditions of a generator determine the material that is used for brushes. **Graphite brushes** are easily distinguished by their silvery appearance and soft flaky texture. This type of brush is for general-purpose use. **Carbon brushes** are used on generators that have both low rotation speeds and a low current output.

Electrographite brushes are made from the same material as carbon brushes but they are processed at high temperatures in an electric furnace. This process increases the ability of the brush to conduct both electrical and heat energy. These brushes are nonabrasive and cooler running. They have very low friction and a higher current capacity than carbon brushes.

Copper-graphite brushes are made from a mixture of powdered copper and powdered graphite pressed together and baked at low temperatures. This type of brush is used on low-voltage generators.

Q2-5. **Normal wear of the commutator segments and the brushes of a generator is compensated for by the _ _ _ _ _ - _ _ _ _ _ _ _ _ _ _ _ _ _ _.**

Q2-6. **_ _ _ _ _ _ _ _ _ _ _ _ _ brushes are used in low-voltage generators.**

THE ARMATURE CORE

Armature coils are wound on cores of either the **Gramme-ring** or **drum** type. The Gramme-ring armature core is an older type and is usually not found on newer machines.

Gramme Ring

In a Gramme-ring armature, the windings are wound on the surface of an iron or steel ring. The armature winding is then tapped at regular intervals for connection to the commutator.

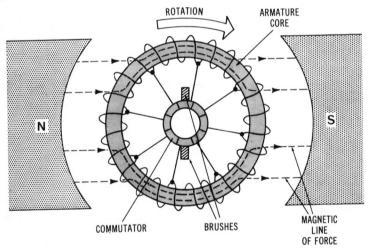

Fig. 2-8. Gramme-ring armature.

Since the armature windings of the Gramme-ring core are wound on the surface of the core, the distance between the core and the pole pieces must be great enough that there is no danger of the armature windings touching the pole pieces. This causes a reduction in the strength of the magnetic field. The weakened field results in a reduced output from the generator.

The Gramme-ring core offers a lower **reluctance** (magnetic equivalent of resistance) to the lines of force than the air gap inside the ring. The magnetic lines of force follow the circular path of the core and do not cross the center air gap. Thus, they are not cut by the conductors on the inside surface of the core and, therefore, do not induce an emf in these conductors.

For these and other reasons, the Gramme-ring armature is rarely used. Nearly all modern dc generators use drum-type armatures.

Drum

The drum armature core is newer and more efficient. Coils are placed in slots and are generally a fraction of an inch below the outer surface of the core. This type of construction

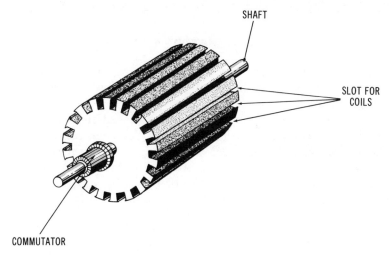

Fig. 2-9. Drum armature core.

and coil placement eliminates the danger of the armature coils rubbing against the pole pieces during the time that the armature is rotating.

Q2-7. What are two disadvantages of the Gramme-ring armature?

Advantages of the Drum Armature Core

One of the advantages of the drum armature core is that a smaller air gap exists between the armature and the pole pieces. The smaller air gap permits the armature coils to cut through a magnetic field of greater density and, thus, a greater emf is induced in the armature coils.

Since the drum armature core is more compact than the Gramme-ring armature core, the drum armature can be ro- tated at a higher speed, thus producing a greater emf.

THE ARMATURE COIL

The coils used on drum armature cores are usually pre- formed. That is, they are wound in their final shape before being put on the armature.

The sides of the preformed coil are placed in the slots of the drum armature core. The two slots for each coil are usu- ally the same distance apart as the adjacent magnetic poles. This distance is called the **pole pitch,** or **pole span.** As you re- call, adjacent magnetic poles are of opposite polarity so that the emf induced in one side of the armature coil is reinforced

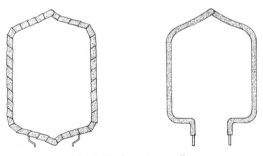

Fig. 2-10. Armature coils.

by the emf induced by the opposite pole on the second side of the armature coil.

Sometimes the distance between the two sides of an armature coil is not equal to the distance between the adjacent magnetic poles. When this distance is less than the pole pitch, the coil is called a **fractional-pitch winding.**

Fractional-pitch windings are used when it is necessary to save copper. The emf induced in a fractional-pitch coil is not as great as the emf induced in a coil whose pitch is equal to the pole pitch. This is because the voltages induced in the two coil sides do not reach their maximum values at the same time.

Lap and Wave Windings

There are two ways the coils can be connected—**lap winding** and **wave winding.** In a **simplex lap winding,** the ends of each coil are connected to adjacent commutator segments. In this way, all the coils are connected in series. This is shown in Fig. 2-11.

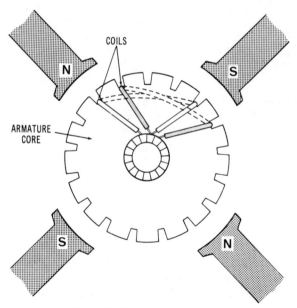

Fig. 2-11. Simplex lap winding.

Q2-8. What are some advantages of the drum armature core?

In a simplex lap winding, a single brush shorts the two ends of a single coil. In a **duplex lap winding,** there are in effect two separate sets of coils, with each set connected in series. The

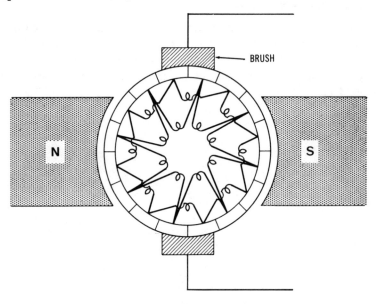

Fig. 2-12. Duplex lap winding.

two sets of coils are connected to each other only by the brushes. Similarly, a **triplex lap winding** is in effect three separate sets of series-connected coils.

In a **wave winding,** the ends of each coil are connected to commutator segments that are two pole spans apart. Instead of shorting a single coil, a brush will short a small group of coils that are in series. There will be as many coils in the group as there are **pairs** of magnetic poles.

Fig. 2-13 illustrates part of a wave winding.

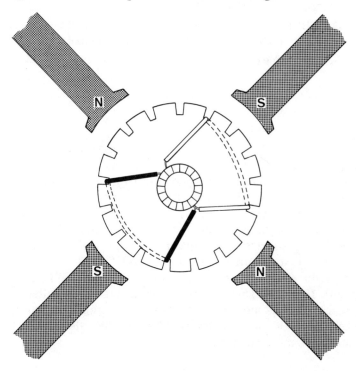

Fig. 2-13. A wave winding.

The Neutral Plane

The area in the generator where no emf can be induced in an armature coil in called the **commutating,** or **neutral, plane.** This plane is midway between adjacent north and south field poles (somewhat shifted from this position under load). In this plane, the moving armature coils cut no lines of force and, therefore, generate no voltage. The brushes are always set so that they short-circuit the armature coils passing through the neutral plane while, at the same time, the output is taken from the other coils.

Q2-9. **In a wave winding, the coil ends are connected to commutator bars two _ _ _ _ _ _ _ _ _ apart.**

Q2-10. **Of what does a duplex winding consist?**

Commutator Output

While the brush is short-circuiting one armature coil, it is receiving the emf and current induced in the other armature coils. This is accomplished by connecting one end of two different coils to the same commutator segment. The illustration in Fig. 2-14 shows an armature with 22 coils connected to 22 commutator segments. There are two brushes. The positive brush is short-circuiting armature coil 11 while the negative brush is short-circuiting armature coil 22. There is no emf being induced in either of these coils. The two coil groups, 1 through 10 and 12 through 21, are connected in parallel by the brushes. This is possible because the voltages in both coil groups have the same polarity. The brushes also connect the generated emf to the load.

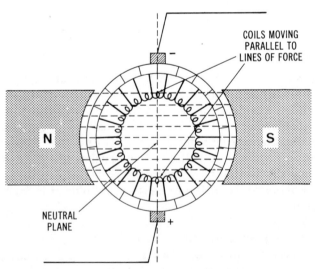

Fig. 2-14. Brush-commutator action in a simplex lap armature.

POLE PIECES

Permanent magnets can only produce magnetic fields of limited strength and are used mainly in small generators called **magnetos.** Most generators use electromagnets to produce the magnetic field. An electromagnet consists of a metal core with a coil of wire wrapped around it.

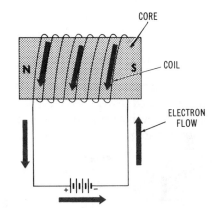

Fig. 2-15. An electromagnet.

When a current flows through the coil, a magnetic field is produced around the coil. If the core conducts the magnetic lines of force easily, it becomes a temporary magnet when current flows through the coil. The current supplied to an electromagnet is dc, since ac would cause the polarity to constantly change.

To find the polarity of the core when current is flowing through a coil, imagine that you wrap the fingers of your left hand around the coil in the direction of the electron flow through the turns. Your thumb then points to the north pole of the core.

The electromagnetic poles are mounted on the frame of a dc generator. This enables the frame to complete the magnetic circuit.

Q2-11. The magnetic field of most electromagnets is (stronger, weaker) than the magnetic field of a permanent magnet.

Q2-12. Why is there no large flow of current in a generator coil that is shorted by a brush?

THE SEPARATELY EXCITED GENERATOR

The current for the electromagnetic field of a generator may be generated by a separate source of direct current. This source could be a battery or a separate dc generator. With either source, the generator is said to be **separately excited.**

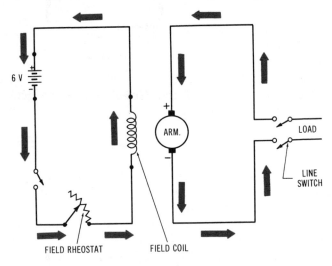

Fig. 2-16. A separately excited generator.

Notice that the separately excited magnetic-field circuit is completely independent of the generator circuit. The strength of the externally excited magnetic field depends directly on the amount of current supplied by the external dc source. A rheostat is used to vary the strength of the magnetic field. This provides a very sensitive control of the generator output.

THE SELF-EXCITED GENERATOR

The current for the electromagnetic field of a dc generator may be developed in the armature coils of the generator itself. The current is taken from the commutator by the brushes and then fed to the field coils. The three types of circuits used to feed current to the field coils are the series, shunt, and compound circuits.

The electromagnetic field of a self-excited dc generator depends on current that is induced into the armature coils of the generator by an electromagnetic field. Since a field is required to produce current, how can a self-excited dc generator build up a voltage?

An electromagnet has a small amount of magnetism even when the electromagnetic coil is not energized. This small amount of magnetism is called **residual magnetism.** Usually, the residual magnetism in the electromagnetic core is strong enough that a weak voltage is induced in the armature coils. This voltage causes a small amount of current to flow in the field windings. The current causes the magnetic field to increase. This increases the voltage which again increases the current, and so on. In this way, the generator voltage builds up to its maximum value. This process is called **building up.**

When there is not enough residual magnetism in the field core, the generator will not build up. In this case, the electromagnetic field must be excited temporarily from an external dc source. This is called **flashing the field.** Reversing the connections to the field windings can also cause a generator to fail to build up.

Q2-13. A generator that has a separate source of current for its field windings is said to be _ _ _ _ _ _ _ _ _ _ _ _ _ _ _ _ _ .

Q2-14. A _ _ _ _ _ _ _ _ is used to control the amount of current flowing in the field winding.

Q2-15. A rheostat provides a (poor, sensitive) control of the output of a generator.

Q2-16. The small amount of magnetism that remains in the core of an electromagnet when the current is removed is called _ _ _ _ _ _ _ _ _ _ _ _ _ _ _ _ _ _ .

The Series Generator

In a series generator (Fig. 2-17), the field windings, armature windings, commutator, brushes, and the external load are all connected in series. In order for a self-excited series generator to produce voltage, the external load must be connected. Without the external load connected, the series circuit is incomplete, and the small generator voltage is due only to residual magnetism.

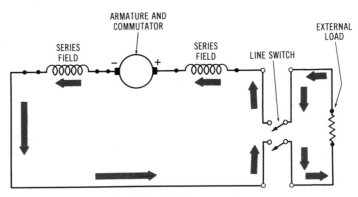

Fig. 2-17. A series generator.

The Shunt Generator

The shunt generator is connected as a parallel (shunt) circuit. The parallel circuit is used to provide separate paths for supplying current to the electromagnetic field and to the external load. In a shunt generator (Fig. 2-18), the dc current induced in the armature coils is taken from the commutator by the brushes. From the brushes, part of the current is supplied to the electromagnetic field windings, while the main current flow is delivered to the external load.

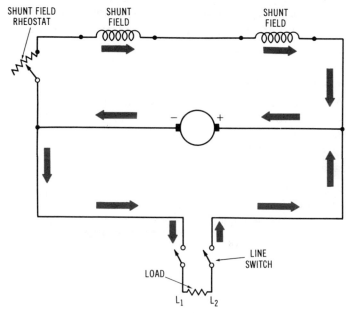

Fig. 2-18. A shunt generator.

Building up in a self-excited shunt generator does not require that the external load be connected to the generator because the current supplied to the electromagnetic field does not flow through the external load. This means the generator can be started, and the full strength of the magnetic field and the induced emf can be reached before the external load is connected.

Q2-17. How will an increase in the current that is drawn by the load affect the current through the field coils of a series generator?

Q2-18. What will be the effect of varying the resistance of the shunt field rheostat?

The Compound Generator

A compound generator is a combination of both a series and shunt generator. It has a series field winding on the pole pieces along with a shunt field winding. The total field strength will thus be increased or decreased as the load increases or decreases.

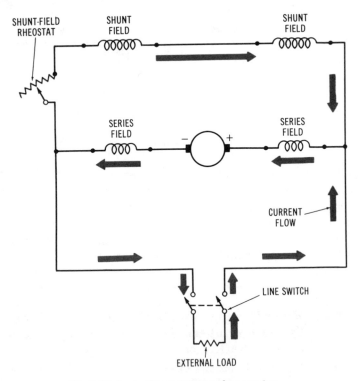

Fig. 2-19. Long-shunt compound generator.

In a **long-shunt** compound generator (Fig. 2-19), the shunt field winding is connected to the ends of the series field windings **farthest away** from the armature. In a **short-shunt** compound generator (Fig. 2-20), the ends of the shunt field winding are connected **between** the series field windings and the armature.

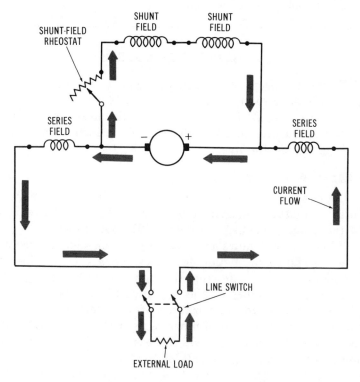

Fig. 2-20. Short-shunt compound generator.

There is no difference between short- and long-shunt compound generators as far as the operation and output are concerned. The only difference is in the point where the series field is connected.

Q2-19. **Compound-generator pole pieces have both _____ and _____ field windings.**

Q2-20. **Name the three types of circuits used to supply current to the field of a self-excited generator.**

GENERATOR LOSSES

There are several effects that take place within a generator. It is necessary to know how these effects are created and how they are corrected in order to maintain electrical equipment in top operating condition. These effects include copper loss, eddy currents, and hysteresis loss.

Copper Loss

Copper loss is due to the resistance of the copper wire used in the armature coils. As current flows through the resistance of the armature coils, there is an I^2R (power) loss.

When current flows through any metal, heat is produced. The resistance of metals increases when their temperature increases. Current flowing through the armature windings causes the resistance of the windings to increase. For example, if the no-load temperature of an armature winding is 68 °F and the full-load temperature is 122 °F, the resistance of the armature winding will increase by about 20% when the temperature rises to the full-load value.

Most generators are constant-voltage devices. The current induced in the armature windings depends on the demands of the external load. As the current demand varies, the copper loss in the armature will vary also.

Eddy Currents

When the conductors of the armature coils rotate through the magnetic field, an emf is induced. Current is caused to flow in the armature coils as determined by the requirements of the external load.

The armature core is also a conductor and it rotates in the same magnetic field. Therefore, an emf and a current are both produced in the armature core. These are called **eddy currents.**

The power used in generating eddy currents comes from the generator power source and represents a loss of output power. This is because it is power taken from the power source but not converted into the desired output; it lowers the efficiency of the generator. (Besides being a waste of power, this energy results in an undesirable heating of the iron.) Eddy currents are reduced by **laminating** the core. That is, the core is made up of a number of thin layers of dynamo sheet steel, all insulated from each other. Lamination reduces the length of the conductor in which the eddy currents flow and, therefore, reduces the amount of I^2R (power) loss in the armature core.

Hysteresis Loss

Hysteresis loss is a heat loss due to the magnetic properties of the armature. This also is a power loss since this energy, too, is taken from the prime mover. When the armature core is rotating, the magnetic particles of the core tend to line up with the magnetic field. Since the core is rotating, the magnetic field keeps changing direction. The movement of the magnetic particles as they keep trying to align themselves produces friction which, in turn, produces heat. This heat results in an increase in armature resistance and an additional copper loss. Hysteresis loss varies with the speed of the armature and with the amount and the type of iron that is used in the core.

To limit hysteresis loss, an armature-core material is used in which the magnetic particles line up with the constantly changing direction of the magnetic field with relative ease. The most commonly used material is dynamo sheet steel. Using this material reduces the hysteresis loss but does not eliminate it completely.

Q2-21. **What two types of power loss are limited by constructing an armature core of dynamo sheet-steel laminations?**

Q2-22. **What happens to the temperature of the armature as the load increases?**

Q2-23. **What effect will this change in temperature have on copper loss in the generator?**

ARMATURE REACTION

Armature reaction is the result of the magnetic field surrounding the armature and is due to the electric current flowing through the armature coils. This field acts on the main magnetic field of the generator and distorts it. This effect is called **cross magnetization.** It only occurs when current is flowing through the armature.

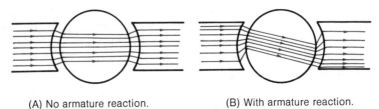

(A) No armature reaction.   (B) With armature reaction.

Fig. 2-21. Distortion of magnetic field.

You have learned that the best place to take the output from an armature is from commutator bars in the **neutral,** or commutating, **plane.** When there is no current flowing through the armature, the neutral plane is at right angles to the main generator field. With armature current flowing, the distorted magnetic field due to cross magnetization will cause the neutral plane to shift to a new position. The position of the neutral plane will shift with every change in current flow in the armature.

Identifying and Correcting Armature Reaction

When sparking occurs between the commutator and the brushes, you should suspect armature reaction. The neutral plane may be found by moving the brushes on the commutator until sparking stops. The neutral plane will, of course, shift each time the current changes.

One way of compensating for armature reaction is to cancel the armature field with an opposing magnetic field. Interpoles and compensating windings are used to provide the needed opposing fields.

Interpoles

Interpoles, or **commutating poles,** are narrow auxiliary poles located midway between adjacent main magnetic poles. Interpoles tend to neutralize the armature reaction created by the magnetic field of the armature. This effect takes place only within the area of the interpole magnetic field.

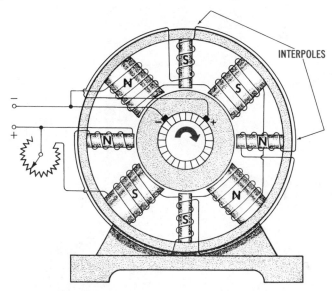

Fig. 2-22. Shunt-field generator with interpoles.

Interpole windings are connected in series with the armature windings and are wound in such a direction that the polarity of the interpole always has the same polarity as the nearest main-field pole in the direction of rotation.

Q2-24. What causes armature reaction?

Q2-25. What symptom indicates armature reaction?

Compensating Windings

When a generator must operate at high efficiency under varying external load demands, **compensating windings** are used. The compensating windings are normally embedded in the

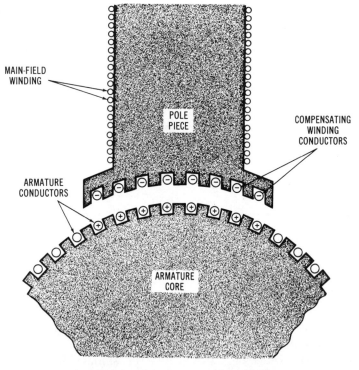

Fig. 2-23. Location of compensating windings.

faces of the pole pieces. They are connected in series with the armature coils, like interpoles, but are wound in the opposite direction from the armature coils. The field of the compensating windings completely cancels the field of the armature that would tend to distort the main field of the generator.

Compensating windings are a very efficient but expensive method for eliminating armature reaction. The use of compensating windings is limited almost exclusively because of their high cost.

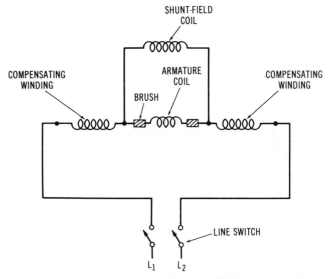

Fig. 2-24. Shunt-field dc generator with compensating windings.

Motor Reaction in a Generator

When current flows in the armature coils, a generator will tend to act as a motor. The magnetic field surrounding the armature coils weakens the main magnetic field on one side of the coil, but strengthens it on the other side. As a result, a torque is developed that opposes the rotation of the generator shaft. This is why the rotating armature presents a mechanical load to the prime mover. Due to this reaction, the prime mover "feels" an opposition to turning the generator shaft.

Q2-26. Why are compensating coils wound in a direction opposite that of the armature coils?

SERIES DC GENERATOR CHARACTERISTICS

To obtain an output voltage, a series dc generator must operate with the line switch closed. This completes the series circuit by including the external load. Without the external load
connected to the series circuit, there is no complete circuit and
the series generator will not build up.

Building Up

When a complete series circuit exists and the armature is
rotating in the magnetic field, an emf is induced in the armature. At first, this emf is due to the residual magnetism present in the iron core of the field poles. This small emf causes a
small load curent to flow through the series field winding. This
causes an increase in the number of magnetic lines of force.
When the number of lines of force generated by the series field
winding increases, the voltage induced in the armature coils
increases.

The output voltage in the series machine continues to rise
until the iron core of the series field becomes **saturated**; that
is, it cannot contain any additional lines of force. When the
magnetic field reaches saturation, the output voltage will be
maximum and any additional current required by the external
load will not increase the output voltage.

Characteristics Under Load

Drawing more current from the armature than that required
to achieve maximum voltage output will create a drop in the
output voltage of the generator. This decrease in voltage is
due to armature reaction and the voltage drop across the resistances of the armature and the field. A graph of the output
of a dc generator is shown in Fig. 2-25.

The most important conclusion that can be reached about the series-wound dc generator is that it is best suited for **constant-current applications.**

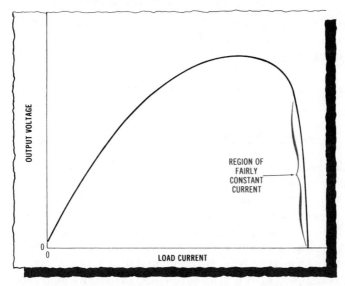

Fig. 2-25. Output-voltage characteristics of a series generator.

As more current is demanded (within practical limits) from a series-wound dc generator, the output voltage increases. When the practical limit of current is reached, a further demand has a reverse effect. The practical limit is reached when the magnetic field strength reaches its maximum value.

Examining the characteristic curve of the series-wound dc generator shows that the voltage drops off sharply. Therefore, it is possible to obtain a fairly constant current for a wide voltage range.

Q2-27. In order to obtain output voltage, a series generator must always be operated with the _ _ _ _ connected.

Q2-28. The series dc generator is best suited for
_ _ _ _ _ _ _ _-_ _ _ _ _ _ _ applications.

Q2-29. The output voltage of a series generator first _ _ _ _ _ _ _ _ _ and, then, _ _ _ _ _ _ _ _ _ as the output current increases.

Your Answers Should Be:

A2-27. In order to obtain output voltage, a series generator must be operated with the **load** connected.

A2-28. The series dc generator is best suited for **constant-current** applications.

A2-29. The output voltage of a series generator first **increases** and, then, **decreases** as the output current increases beyond core saturation.

SHUNT DC GENERATOR CHARACTERISTICS

The field coils of a shunt-wound dc generator are connected in parallel with the armature and the external load. The number of magnetic lines of force produced by the pole pieces does not depend directly on the load current. However, the load current does have an indirect effect. As the load current increases, the generator output voltage decreases. This is due to increased armature reaction and the voltage drop across the resistance of the armature coils. When the generator output voltage decreases, the current through the shunt field coils also decreases. This causes an additional decrease in the output voltage. The output voltage is, therefore, less than it would be if the field windings were connected to a source of constant voltage.

A rheostat is usually placed in series with the shunt field windings. By adjusting the rheostat, the current through the windings can be controlled. In this way, the strength of the magnetic field can be changed to make up for a decrease in output voltage.

Building Up

The line switch to a shunt dc generator can be left open since it is not necessary for the external load to be connected during buildup. When the generator is turning, the armature coils cut the weak residual magnetic field of the iron shunt-field core. The weak magnetic field causes a weak voltage to be induced in the armature coils. This voltage, in turn, causes a weak current to flow from the armature coils, through the shunt-field rheostat, and through the shunt-field coils. The weak shunt-field current generates additional lines of force which strengthen the magnetic field. The strength of the electromagnetic field and the voltage induced in the armature coils increase until the terminal voltage of the generator reaches its no-load voltage.

Characteristics Under Load

In order for the shunt dc generator to deliver power to the external load, the line switch is closed. The load requirements are met by adjusting the shunt-field rheostat.

As the amount of current required by the external load increases, the output voltage of the generator decreases as shown by the characteristic curves of Fig. 2-26.

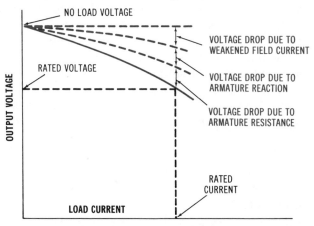

Fig. 2-26. Output-voltage characteristics of a shunt generator.

The shunt dc generator is well suited for **constant-voltage applications** at a specified rated output. When an additional load (beyond a critical limit) is placed on the generator, the output voltage falls off almost to zero. Any attempt to force a shunt dc generator to deliver more than its rated output could cause it to break down.

Q2-30. A shunt dc generator is used mainly for constant _ _ _ _ _ _ _ applications.

Q2-31. When more current is required by the external load than the critical limit of a shunt generator, the additional load will cause the output voltage to _ _ _ _ .

COMPOUND DC GENERATOR CHARACTERISTICS

You have learned that there are two types of compound dc generators—the long shunt and the short shunt. The buildup, loading, and general characteristics of the two are very similar. The short-shunt generator is in wider use because of its simpler circuitry. On the next few pages, the compound dc generator will be discussed in terms of the short-shunt design. The series field of a short-shunt compound dc generator is connected between the load and the parallel shunt field and armature. Basically, the compound dc generator takes advantage of the characteristics of both the series and shunt dc types.

Building Up

One of the advantages of the shunt generator is also present in the compound generator. It is not necessary to start a compound generator with the external load connected. The building-up process is similar to that of the shunt type.

Characteristics Under Load

When the no-load terminal voltage has been reached, the line switch is closed. The output voltage is then adjusted with the shunt-field rheostat which controls the resistance of the shunt-field circuit. As described before, this varies the strength of the magnetic field. The output voltage changes directly with the strength of the magnetic field that is controlled by the shunt-field rheostat.

The purpose of the series field windings in a compound generator is to control the output voltage of the generator in relation to the external load. The series field windings help to offset the voltage decrease that occurs when a shunt field alone is used.

Compounding Effects

When the effect of the series winding produces the same terminal voltage at rated load as at no load, the generator is said to be **flat compounded**.

When the effect of the series winding produces a smaller terminal voltage at rated load than at no load, the generator is said to be **undercompounded**.

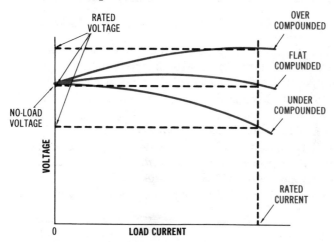

Fig. 2-27. Compound-generator characteristics.

When the effect of the series winding produces a greater terminal voltage at rated load than at no load, the generator is said to be **overcompounded**. Compound generators are usually wound so that they are slightly overcompounded.

In some generators, the degree of compounding is controlled by a variable resistor in parallel with the series field. This resistor is called a **diverter**. It determines the fraction of the load current that flows through the series field. This gives the same effect as changing the number of turns in the winding.

Q2-32. A compound generator most resembles a _ _ _ _ _ generator in its operation.

Q2-33. A compound generator in which the rated-load voltage is equal to the no-load voltage is said to be _ _ _ _ _ _ _ _ _ _ _ _ _ _.

AUTOMATIC VOLTAGE REGULATION

It is possible to use an automatic device to keep the output voltage of a shunt generator nearly constant, even if the load changes. Such a device is called a **voltage regulator.** These regulators automatically change the current through the shunt-field winding every time the output voltage starts to vary.

Some regulators have a variable resistance in series with the field winding. If the output voltage decreases, the resistance is decreased, and the voltage is brought back almost to its original value. If the voltage increases, the resistance increases, and the voltage again is returned almost to normal.

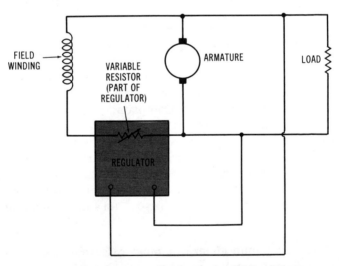

Fig. 2-28. Voltage regulation by varying field resistance.

In another type of regulator, a pair of contact points is used to short-circuit a fixed resistor in the field circuit. When the contacts are open, the resistor limits the field current and keeps the output voltage below the desired value. When the contacts short-circuit the resistor, the voltage rises above the desired value. In operation, the regulator causes the contacts to vibrate (open and close rapidly). The average current through the field winding depends on how rapidly the contacts vibrate. This, in turn, is determined by the value of the output voltage.

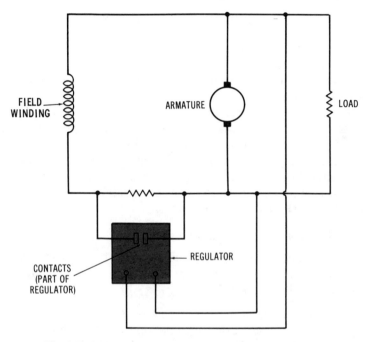

Fig. 2-29. Voltage regulation by using vibrating contacts.

More complicated arrangements are sometimes used to regulate the voltage of shunt generators. However, they all operate by controlling the amount of current flow in the shunt-field winding.

Q2-34. The voltage of a shunt generator is regulated by varying the _ _ _ _ _ - _ _ _ _ _ current.

Q2-35. A device that automatically keeps the output voltage of a generator nearly constant is called a

_ _ _ _ _ _ _ _ _ _ _ _ _ _ _ _ _.

PARALLEL OPERATION OF DC GENERATORS

At times, it is necessary for more than one generator to supply electrical energy to the same load. This may be due to peak-load demands or the need for continuous service should one generator become disabled.

In order to have more than one generator supplying the external load, it is necessary to connect them in parallel. The same general precautions should be taken when connecting dc generators in parallel as are used when connecting batteries in parallel. Polarity and voltage must be the same. It is important to remember that **different paralleling procedures are used for ac than for dc generators.**

Paralleling Two Dc Generators

Fig. 2-30 shows two generators that can be connected to the same load by means of switches. The procedure for connecting them in parallel is as follows.

1. Generator G_1 is supplying the external load; generator G_2 is to be placed into parallel operation with G_1.
2. Switch S_2 must be open to prevent generator 1 from trying to operate generator 2 as a motor.
3. Adjust the shunt-field rheostat of generator 2 to the lowest position (maximum resistance).
4. Bring generator 2 up to its rated speed.
5. Adjust the shunt-field rheostat so that generator 2 is supplying slightly more voltage than generator 1. The polarity must be as shown.
6. Close switch S_2 to bring generator 2 into parallel operation with generator 1. NOTE: Generator 2 should be carrying a small portion of the external load.
7. Adjust the shunt-field rheostat of generator 2 to distrib-

ute the load equally. At the same time, the shunt-field rheostat of generator 1 is adjusted to maintain the normal voltage.

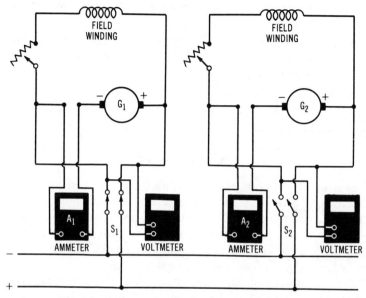

Fig. 2-30. Two dc generators connected in parallel.

Removing a Dc Generator From Parallel Operation

To remove a dc generator from parallel operation:
1. Weaken the field of the generator that is to be removed from operation and, at the same time, strengthen the field of the remaining generator.
2. When the outgoing generator is no longer carrying any of the load, open the switch so that the generator is removed from operation.

Q2-36. Are the same procedures used in paralleling ac and dc generators?

Q2-37. When paralleling dc generators, the positive terminal of one generator must be connected to the _____ terminal of the other.

Q2-38. The load is transferred from one generator to the other by adjusting the _____ - _____ _____.

MAINTENANCE

In order to operate any machine properly, you must be familiar with the construction details and maintenance procedures for the machine. Maintenance should not be confined to the repair and replacement of units that have failed in operation. Regular checks on operating equipment aid in detecting many problems before they become serious.

The nameplate on the generator should list the **maximum temperature rise** for the machine. This is the maximum amount by which the machine should be warmer than the surrounding air.

Check for generator hot spots by touching it with the palm of your hand. Check for faulty bearing operation by feeling the bearing caps on the end bells. Naturally, you must be careful to avoid injury when touching the generator. Table 2-1 lists some causes and remedies of hot spots.

When the bearings are checked for overheating, you should also test for any vibration. Vibration may be caused by excessive speed of the rotating shaft and is usually quite loud. If not corrected, this can result in permanent damage to the machine. Excessive speed, if maintained for any length of time, can result in the machine tearing itself apart. Vibration may also be caused by poorly balanced rotating parts or by worn bearings.

It is important to follow the manufacturer's instructions for the lubrication and maintenance of bearings. This information will usually be contained in the manual that accompanies the equipment. **Follow the manufacturer's instructions exactly.**

Pointers About Lubrication

Remember these things about lubrication. Improper lubrication, either too little or too much, can cause serious damage to moving and electrical parts. Too little lubrication causes friction and wear of moving parts. Excessive lubrication can cause electrical damage—shorting the commutator segments, fouling the commutator brush assembly, or soaking the armature coils.

Table 2-1. Causes and Remedies of Generator Hot Spots

Possible Causes	Remedy
1. Insufficient lubrication.	1. Lubricate.
2. Excessive load on generator.	2. Check ammeter reading with current rating of the generator. If overloaded: A. Reduce generator load. B. Place a second generator in parallel to share the load.
3. Clogged cooling vents.	3. With generator off, blow out with clean dry air at a low pressure.

Q2-39. How would you determine the allowable temperature rise for a generator?

Q2-40. What could cause a generator bearing to become too hot?

Q2-41. What are some causes of vibration in a generator?

WHAT YOU HAVE LEARNED

1. The main parts of a dc generator are the frame, end bells, pole pieces, shaft, armature, commutator, and brushes.
2. The armature contains a number of coils in order to produce a relatively smooth output.
3. Various types of carbon and graphite brushes are used to take voltage and current from the commutator segments.
4. Most armatures consist of a drum-type iron core with windings set in slots in the core.
5. Armature coils are connected to the commutator in simplex, duplex, or triplex patterns.
6. Coils can be connected in a lap-wound or wave-wound pattern. In a lap-wound pattern, a single coil is shorted by a brush; in a wave-wound pattern, a small group of coils in series is shorted.
7. The neutral plane is the plane where no emf is induced in a coil. The brushes should be set to this position.
8. The electromagnetic field of a dc generator can be either separately excited by an external dc source or self-excited from the generator output in a series, shunt, or compound circuit.
9. A self-excited generator can start from its own residual magnetism or, if necessary, it can be started by flashing the field with an external dc source.
10. The current flow in a series generator varies with the current required by the external load and causes the strength of the magnetic field to vary.

11. Armature reaction results in the shifting of the neutral plane due to interaction of the magnetic field caused by current flow in the armature with the main magnetic field.

12. A series generator and its external load must be connected in order to obtain voltage from the generator.

13. Generator losses are copper loss (due to armature resistance), eddy-current loss, and hysteresis loss.

14. The series generator is best suited for constant-current applications.

15. The shunt generator does not require that the external load be connected during buildup of that generator.

16. The no-load terminal voltage of a shunt generator is greater than the rated voltage (voltage after external load is connected) because the load-current flow causes power losses within the generator.

17. The current supplied by a shunt generator varies with the external load requirements, but a fairly constant voltage is maintained.

18. The compound generator is basically a shunt generator with a series field that is used to offset the falling voltage when large amounts of current are required by the external load.

19. The ability of the series field of a compound generator to produce a greater, equal, or lesser terminal voltage than the no-load voltage determines if the generator is over, flat, or undercompounded.

20. Regulators can be used to keep the output voltage of a shunt generator nearly constant.

21. A shunt generator can be connected in parallel with another shunt generator only when it has reached a voltage slightly above the voltage of the other generator.

22. Generators should be checked regularly for hot spots and vibration.

23. It is important to follow the manufacturer's instructions for the lubrication and maintenance of generators.

3

DC Motors

In this chapter, you will learn about dc motors. You will see how they are constructed and find out what reactions occur inside them. You will be able to recognize the different types of dc motors and know the advantages and disadvantages of each. You will learn about starting devices for dc motors, why they are needed, how they work, and the various methods used to control the speed of dc motors. You will also learn about dc brushless motors and methods of reducing interference caused by motors.

BASIC DC MOTOR CONNECTIONS

The parts of a dc motor are essentially the same as the parts of a dc generator. A dc motor consists of a frame, end bells, pole pieces, shaft, armature assembly, commutator assembly, and brush assembly. The construction of these parts is essentially the same as for the generator, with minor changes for practical reasons.

The field windings on the pole pieces (also called the **stator**) are supplied with direct current. The commutator and armature (sometimes called the **rotor**) are also supplied with dc which is converted to ac, as you learned in Chapter 1. It would be possible, of course, to supply the two windings from different dc sources, but normally they are both supplied from the same source.

The two windings (stator and rotor) then become parts of the same circuit. The methods of connecting the two windings determine the different types of dc motors. The three basic types are **shunt, series,** and **compound** motors.

Notice that the dc motors in the diagrams of Fig. 3-1 are similar to the three types of dc generators.

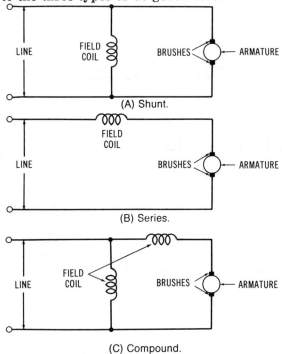

(A) Shunt.

(B) Series.

(C) Compound.

Fig. 3-1. Types of dc motors.

The direction of rotation of any dc motor can be reversed by reversing the leads of the field coils but leaving the armature connections unchanged. If both the field coil and the armature leads are reversed at the same time, the motor will continue to run in the same direction.

Fig. 3-2. Effect of reversing field connections in a dc motor.

Reversing the leads of the field coils reverses the polarity of the magnetic field. The armature field remains unchanged. This causes the motor to run in the opposite direction, as shown in Fig. 3-2. The magnetic poles of the rotor and stator, as shown, attract each other until the field connections are reversed. Then, the poles repel each other. If both the field and armature connections were reversed, the poles would continue to attract each other, and the motor would not reverse direction.

ARMATURE LOSSES

Armature losses occur in a dc motor for exactly the same reasons they occur in a dc generator. The losses in the armature of a dc motor are copper loss, hysteresis loss, and eddy-current loss. The same kinds of construction are used in dc motors to reduce the armature losses as in dc generators. Armature cores in motors are usually built of laminated dynamo sheet steel.

ARMATURE REACTION

Just as in a generator, the interaction of the armature and the main magnetic fields distorts the main field. This causes the commutating plane of the motor to shift. There will be excessive sparking when the brushes are not properly aligned in the commutating plane. Armature reaction in the motor can be limited by interpoles and compensating windings.

Q3-1. What happens to the direction of rotation of a dc motor if the leads of the field, but not of the armature, are reversed?

Q3-2. There are armature losses in a dc generator. Are there similar losses of power in a dc motor?

Q3-3. Armature reaction occurs in a dc generator. Is there armature reaction in a dc motor?

Q3-4. Another name for the stationary field assembly in a motor is the _ _ _ _ _ _ .

Q3-5. Another name for the rotating armature is the _ _ _ _ _ .

COUNTER EMF

When the armature conductors of a motor rotate in the main
magnetic field, an emf is generated. This always happens when
a conductor cuts magnetic lines of force. This emf opposes the
applied line voltage. The faster the motor turns, the greater
the **counter emf** becomes.

When a dc motor is started, a very large current will flow
unless a starting resistor is used to limit this current. As the
motor builds up speed, however, the counter emf increases and
limits the current by reducing the effective voltage across the
armature coils.

DC SHUNT MOTORS

The shunt-field winding consists of many turns of small wire
and is connected in parallel with the armature winding, or
across the line, as shown in Fig. 3-3.

The shunt-type motor is used when it is desired to vary the
rotational speed above and below the normal speed (Fig. 3-4).
Increasing the resistance in series with the shunt field will

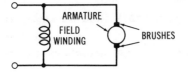

Fig. 3-3. A dc shunt motor.

cause the motor to speed up. (If the field circuit is broken and voltage is still applied to the armature, the motor may run so fast that it damages itself.) A resistor connected in series with the armature will decrease the speed of the motor. The characteristics of any type of motor need to be known so that the

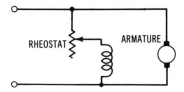

Fig. 3-4. A dc shunt motor with field rheostat.

motor may be used properly. Fig. 3-5 indicates some typical characteristics for a shunt motor.

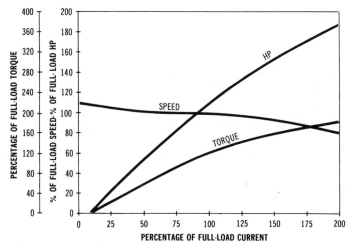

Fig. 3-5. Typical shunt-motor characteristic curves.

The main characteristics of a shunt motor are its constant speed and low starting torque. This means that dc shunt motors cannot be used for hard-starting loads.

Q3-6. The counter emf _ _ _ _ _ _ _ _ when armature speed increases.

Q3-7. What are the two main characteristics of a dc shunt motor?

Q3-8. A resistance added in series with the shunt field _ _ _ _ _ _ _ _ the speed of the motor.

DC SERIES MOTORS

A dc series motor has its field and armature windings connected in series. Both sets of windings, therefore, carry the same current.

The field coil, therefore, must be of much heavier construction than the field coil of a dc shunt motor in order to withstand the heavy currents. The schematic diagram of a dc series motor is shown in Fig. 3-6.

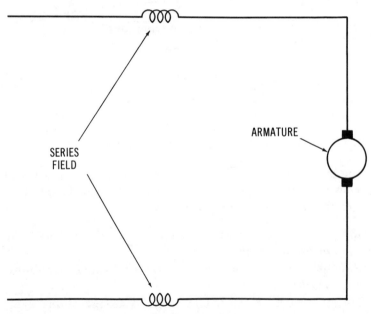

SERIES FIELD

ARMATURE

Fig. 3-6. A dc series motor.

The starting and stalling torque of a series motor is excellent (Fig. 3-7). It will start and carry very heavy overloads. Its speed regulation, however, is very poor. The speed of a series motor decreases as the load increases. Therefore, the

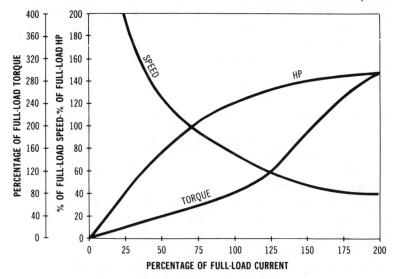

Fig. 3-7. Typical series-motor characteristic curves.

motor turns slower with a heavy load and faster with a light load. If a dc series motor is run without any load, it may run so fast that it may damage itself. Series motors are seldom used in applications that require belt coupling to the load; if the belt should break, the motor would be without a load. A typical use of a series motor is the automobile starter.

Q3-9. Why is the field winding of a series motor wound with larger wire than the field winding of a shunt motor?

Q3-10. A series motor has a (large, small) starting torque.

Q3-11. What happens to the speed of a series dc motor that is operated with no load connected to its shaft?

Q3-12. The armature current of a series motor (does, does not) flow through the field windings.

DC COMPOUND MOTORS

As the name implies, the dc compound motor is a combination of a dc shunt motor and a dc series motor. It has both a series and a shunt field winding. Its characteristics are a combination of the characteristics of the dc series motor and the dc shunt motor. When the shunt field coil and the series field coil act to aid each other, the machine is said to be a **cumulative-compound** motor. When the shunt field coil and the series field coil oppose each other, the machine is called a **differential-compound** motor.

Compound motors are divided into two groups according to the manner in which the series field coil is connected with respect to the shunt field coil. These two groups are known as **short-shunt** and **long-shunt** compound motors (Fig. 3-8).

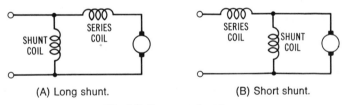

(A) Long shunt.   (B) Short shunt.

Fig. 3-8. Compound motors.

By proper control of the relative strength of the two field coils, the compound motor may be used to meet any requirement that can be met by a pure series motor or a pure shunt motor. The characteristics shown in Fig. 3-9 apply to a typical compound motor, but not necessarily to every compound motor.

Examination of the graph in Fig. 3-9 reveals that the speed of a compound motor is almost as constant as that of a shunt motor and the starting torque is almost as high as that of a series motor. Also, a compound motor does not run dangerously fast even at no load. Compound motors can be designed to have different characteristics.

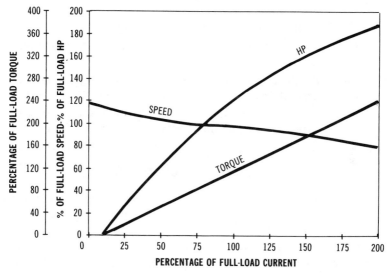

Fig. 3-9. Typical compound-motor characteristic curves.

The compound motor has two field windings. For this reason, a compound motor is usually more expensive than either a series or a shunt motor of the same capacity.

Q3-13. What type of dc motor has the highest starting torque?

Q3-14. What type of dc motor should never be started without a mechanical load connected?

Q3-15. What type of dc motor has a constant speed characteristic?

Q3-16. Shunt dc motors have a _ _ _ starting torque.

Q3-17. The same current flows through the armature and the field coils in a dc _ _ _ _ _ _ motor.

Q3-18. How does a short-shunt compound motor differ from a long-shunt compound motor?

REVERSAL OF DC MOTORS

The direction in which a dc motor rotates can be reversed by either of two methods. One way is to reverse the connections to the armature. The other is to reverse the connections to the field windings. In a series dc motor, it doesn't matter which set of connections—armature or field—is reversed. However, opening the circuit, to reverse connections in a series motor, opens both the armature and the field. Fig. 3-10 shows the basic circuit for the reversal of a series motor. At the left (Fig. 3-10A), the motor might be considered as rotating in

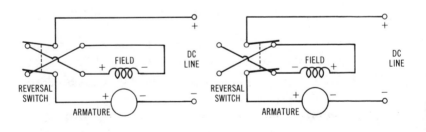

(A) Forward.   (B) Reverse.

Fig. 3-10. Basic method of reversing a series motor.

the forward direction, while in Fig. 3-10B, with the switch thrown the opposite way, the polarity of the field reverses causing the motor to turn the opposite way.

For shunt motors, equipment designers must decide whether to reverse the armature or the field in order to reverse the motor rotation. Actually, in most cases, it's better to reverse the armature connections because there are some problems associated with reversing a shunt field. The only advantage of choosing to reverse the field is that the field current is normally much lower than the armature current and, therefore, low-current switches can be used. Some disadvantages are: (1) if the shunt field is accidentally left open, motor speed may increase greatly and cause injury or damage, (2) use of a switch or relay in the field circuit increases the risk of a poor contact or an open field (again possibly causing damage or

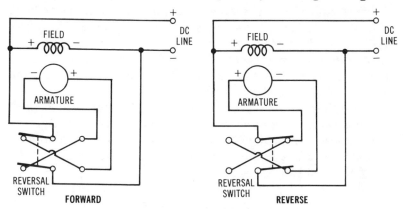

Fig. 3-11. Basic method of reversing a shunt motor.

Q3-19. Reversing the rotation of a series motor can be done either by reversing the _ _ _ _ _ _ _ _ or by reversing the _ _ _ _ _.

Q3-20. Normally, to reverse the rotation of a shunt motor, it is the connections to the _ _ _ _ _ _ _ _ that are reversed.

Q3-21. To reverse the direction of a compound motor, you would reverse the _ _ _ _ _ _ _ _ connections.

Q3-22. A much used method of stopping a motor quickly is dynamic braking. It involves connecting a resistance across the _ _ _ _ _ _ _ _.

injury), and (3) the reversal of the field connections requires the use of a field discharge resistor to prevent excessive voltage buildup at the instant of reversal. In addition, reversing the field connections could interfere with a dynamic braking circuit if one is used. A basic circuit for the reversal of the armature of a motor is shown in Fig. 3-11.

Reversing the direction of a compound motor is always accomplished by reversing the connections to the armature because there is a field coil in series with the armature and the armature current is, therefore, low. Thus, low-current switches or contactors can be used.

As you might already have suspected, reversing the direction of a motor, especially a large one, does not happen instantly. The motor needs time to slow down and stop before rotating in the opposite direction. If a motor is simply disconnected from its source of power, inertia will keep it rotating until it's stopped by friction.

One often-used method for slowing down a motor is to use dynamic braking. That is, connect the armature across a resistance. The motor then acts as a generator with the resistance as its load, thus, slowing the motor more quickly than it would without the resistor across the armature. Normally, the starting resistance serves double duty by also acting as the resistance for dynamic braking. A second way to stop a motor quickly is to reverse the power-line connections; this method is called **plugging.**

MANUAL STARTERS

A dc motor presents a very low resistance to a voltage source whenever the armature is at rest. For instance, a motor that normally draws 50 amperes at full load may have a starting current of 500 amperes or more. This in-rush current could easily damage the motor. In order to prevent this condition, **starters** are often used. Starters are usually required only with

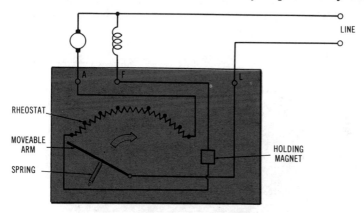

Fig. 3-12. Three-point starting box.

machines that are rated at more than ¼ horsepower. In most cases, the starters are rheostats that are connected in series with the armature. These rheostats are usually contained in an enclosure called a **starting box.**

In a **three-point starting box,** the connections are made as shown in Fig. 3-12. When the voltage is turned on, the rheostat that is in series with the armature limits the armature current to a reasonable level. As the motor gains speed, the operator turns the rheostat arm in the direction of the arrow, one step at a time. Some of the resistance is also in series with the shunt field, but it is small enough to be neglected. The holding magnet is energized by the field current so that when the arm has been pushed across the entire rheostat, the holding

Q3-23. How many external connections are there on a three-point starting box?

Q3-24. When is the arm on a four-point starting box released?

magnet will hold it in this position. If the field circuit should become open, the arm is released and returns to the off position.

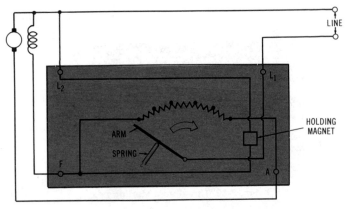

Fig. 3-13. Four-point starting box.

The operation of a **four-point starting box** (Fig. 3-13) is similar to that of the three-point starting box. However, the holding magnet releases the arm when the voltage, rather than the field current, is lost. The four-point box is the more common unit.

A common form of the manual starter is known as a faceplate starter. The faceplate starter receives its name from the fact that most of its contacts and terminals are located on the front or faceplate. Also, its on/off handle, or switch, is located on and operated from the front of the unit. Faceplate starters are used with dc motors of up to about 50 hp.

AUTOMATIC STARTERS

Automatic starters are commonly used when the frequent starting of dc machines is necessary. These starters can be operated by relatively inexperienced personnel since they involve only button pushing. The motor in Fig. 3-14 is a compound motor, but the circuit would apply equally well to shunt motors (by bypassing the series-field coil) or to series motors (by bypassing the shunt-field loop).

As the starting push button is depressed, line voltage is applied to coil M. The coil closes the two normally open contacts, M_1 and M_2. Contact M_1 causes continued current flow through coil M. Contact M_2 causes the armature to receive voltage. The three series resistors R_1, R_2, and R_3 limit the armature current to a safe value. Relays A, B, and C (whose coils are not shown) are operated in order as the counter emf in the armature increases.

Contact A closes first, thus short-circuiting R_1 and leaving less resistance in the armature circuit. Contact B closes next and short-circuits R_2. Finally, contact C closes. This leaves no external resistance in the armature circuit.

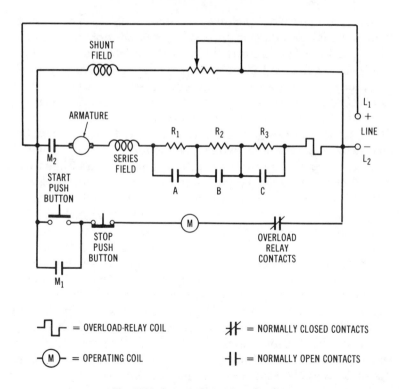

Fig. 3-14. Automatic starter circuit.

When the stop button is pressed, coil **M** is deenergized, and contacts M_1 and M_2 both open. As the motor comes to a stop, contacts A, B, and C open.

If an overload occurs during normal motor operation, the heavy current will activate the overload relay and open its contacts. This is equivalent to pushing the stop button.

The short-circuiting of the resistances in series with the armature can be accomplished by time-delay relays, voltage-sensitive relays, or current-sensitive relays. The theory always remains the same—use series resistance to start and, then, lower the resistance value as the motor picks up speed. Voltage-sensitive relays will operate as the **counter emf** in the armature **increases**. Current-sensitive relays will operate as the **armature current decreases**.

DRUM CONTROLLER

If a motor starter includes features such as speed control and reversal, in addition to starting and stopping, it is often referred to as a **controller**. Faceplate starters and faceplate controllers are used for motors up to about 50 hp, as previously mentioned. Above 50 hp, **drum controllers** are used. A drum controller provides the same functions as a faceplate controller. However, the construction of a drum controller is more rugged, its contacts are heavier and better insulated, and its resistors are usually larger and externally mounted. Also, the handle of a drum controller can normally be operated in either direction. Thus, if it is moved from the forward position backward through the OFF position, the motor will reverse itself. An example of a drum controller circuit is shown in Fig. 3-15.

A controller is designed to accommodate a compound motor. One is shown connected to the controller in Fig. 3-15. The armature, the series field, and the shunt field receive power from the dc line, usually through fuses and a switch or through a circuit breaker. Speed regulation is accomplished through control of both the armature and the field in this system. Also included in the system are means for starting, braking, and reversing. The dynamic method of braking that was mentioned earlier is used here—the armature is connected across the resistor to provide a load for the motor, thus causing it to stop quickly. Rugged fixed contacts in the drum are connected to

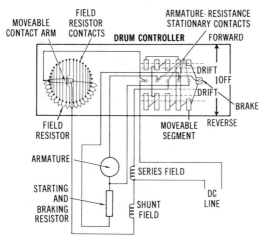

Fig. 3-15. Reversing drum controller.

different resistances when the controller handle is turned and the moveable contacts slide along the fixed contacts to vary the armature resistances. The field resistance is varied when the contact arm moves over the contacts of the field resistors.

Each position of the controller handle provides different steps of speed variation. Forward and reverse operation are indicated by arrows. The OFF position of the controller is in the center position. Moving the handle to the OFF position provides dynamic braking of the motor while the DRIFT position allows stopping without braking.

Drum controllers are normally totally enclosed except for the handle and because they usually operate high horsepower dc motors, they are better insulated. When protection against an excessive load and a low operating voltage is advisable, a separate line starter is used.

Q3-25. What is the purpose of dc motor starters?

Q3-26. Name the three types of relays that can be used to short-circuit resistors in an automatic starter.

Q3-27. Voltage-sensitive relays respond to the _ _ _ _ _ _ _ _ _ _ in the armature.

Q3-28. In a motor starter, current-sensitive relays act when the current _ _ _ _ _ _ _ _ _ .

Q3-29. Under what conditons are automatic starters commonly used?

Some motor controllers include a **jogging** feature. A push-button switch, marked JOG, allows a quick opening and closing of the motor-operating circuit, thus permitting very short periods of rotation of the motor. Jogging is provided for when a careful positioning of a machine is required, either for normal operation or during troubleshooting and maintenance.

RELAYS

There are several types of relays used in dc-motor controllers. The simplest is a switch closed or opened by an electromagnet when the current through the relay coil reaches a certain level (Fig. 3-16).

It is often desirable for a relay to operate only after a cer-

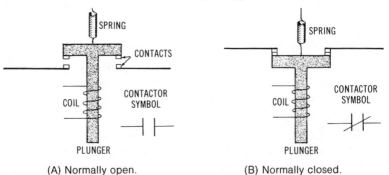

(A) Normally open.   (B) Normally closed.

Fig. 3-16. Simple relays.

tain time delay. For example, the automatic starter described earlier may use time-delay relays to assure that the motor reaches its full load current after a given time, no matter what the load on the motor. Or, it may not be desirable to have an overload relay disconnect the motor as a result of a momentary overload. A time-delay relay can be used for this purpose also.

Thermal Time Delay

Several types of relays can offer a time delay. One type often used for overload protection is the **thermal overload relay** (Fig. 3-17). This relay is operated by heat rather than magnetism. A strip of brass and a strip of copper are fastened together to form a **bimetal strip.** When the bimetal strip is

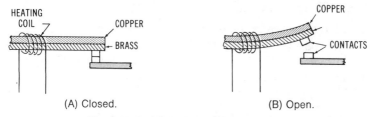

(A) Closed.   (B) Open.

Fig. 3-17. A simple thermal overload relay.

heated, the brass expands more than the copper, and the strip bends away from the fixed contact and opens the circuit. The action of this relay depends on the heating effect of the current. That is, it depends on both the amount and the duration of the current. If a heavy current flows for a short time but not long enough to cause an overheating of the motor, the relay will stay closed.

Q3-30. Drum controllers are used for operating motors rated at more than __ hp.

Q3-31. Speed control of the motor shown in Fig. 3-15 is through control of the armature and the _ _ _ _ _ .

Q3-32. The method used by the drum controller shown in Fig. 3-13 to slow the motor quickly is _ _ _ _ _ _ _ braking.

Q3-33. The push-button control labeled _ _ _ is used to allow a careful positioning of a motor-driven machine.

Dashpot Time Delay

Automatic starters generally use magnetic time-delay relays. However, one type, the **dashpot timing relay,** uses a cylinder containing oil (called a dashpot) to slow down the motion of a plunger (Fig. 3-18).

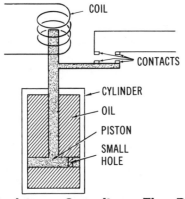

Fig. 3-18. A dashpot relay.

Resistance-Capacitance Time Delay

It is possible to make use of the RC time constant of a resistor and a capacitor to establish a time delay. For practical purposes, a current strong enough to operate a relay will flow from a discharging capacitor for a period that is equal to about five time constants.

In the simplified circuit shown in Fig. 3-19, switch S_1 is closed and switch S_2 is open when the motor is off. The capacitor becomes charged. When the motor is started, switch S_1 is opened and S_2 is closed. The capacitor discharges through re-

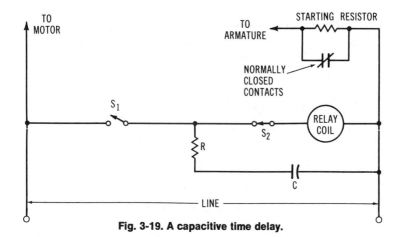

Fig. 3-19. A capacitive time delay.

sistor R and the relay coil. During the first moments after the motor is turned on, enough current flows through the relay to hold its contacts open. This causes the starting resistor to limit the armature current. When the capacitor-discharge current decreases below a certain value, the relay contacts close, short-circuiting the starting resistor. Armature current is no longer limited except by the counter emf which is now high enough to protect the armature.

MOTOR EFFICIENCY

Before motor efficiency can be discussed, the term efficiency must be thoroughly understood. **Efficiency** is the ratio of the amount of power obtained from a machine to the amount of input power required to operate it. Therefore, efficiency is a ratio of the power output to the power input of a system. This statement can be written as follows:

$$\text{Eff} = \frac{P_o}{P_i}$$

If a system requires 1000 watts at the input while delivering 800 watts at the output, its efficiency is:

$$\text{Eff} = \frac{P_o}{P_i} = \frac{800}{1000} = 0.80, \text{ or } 80\%$$

Efficiency is usually expressed as a percentage. In a motor, the output (mechanical power) is measured in horsepower (hp) while the input (electrical power) is measured in watts. The output is multiplied by 746 in order to obtain its equivalent in watts. For example, 1 hp = 746 W, so 2 hp = 2 × 746 = 1492 W. To find the efficiency of the motor shown in Fig. 3-20,

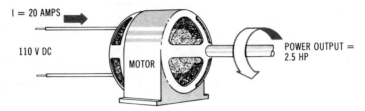

I = 20 AMPS

110 V DC

MOTOR

POWER OUTPUT = 2.5 HP

Fig. 3-20. Calculation of motor efficiency.

you must calculate the power input (in watts) and power output (in watts) and, then, use the formula to find efficiency.

$$P_i = E \times I = 110 \times 20 = 2200 \text{ W}$$
$$P_o = hp \times 746 = 2.5 \times 746 = 1865 \text{ W}$$

Therefore,

$$\text{Eff} = \frac{P_o}{P_i} = \frac{1865}{2200} = 0.85, \text{ or } 85\%$$

Motor Losses That Affect Efficiency

Now that you know how to calculate efficiency, the question of why the output power is smaller than the input power should be answered. A motor has several electrical losses. There are armature losses, losses in the field windings, and losses in the shunt-field rheostat.

There are also mechanical losses in the form of friction. The armature runs on bearings, and the brushes "rub" against the commutator. In addition, the entire armature has to overcome air friction while spinning. The power necessary to provide for these friction losses must be supplied from the input source. A fan is connected to one end of the armature to cool some motors. This amounts to an additional loss due to air friction. These mechanical losses depend mainly on the speed.

Motor efficiency increases as the physical size of the motor increases. For fractional-horsepower motors, efficiencies are about 40–50%. In the 10-hp range, the efficiency is about 85%. This is because the mechanical losses do not increase with the motor size at the same rate as does power output.

SPEED CONTROL

One of the great advantages of dc motors over ac motors is that the speed of dc motors can be controlled easily. There are three basic methods for speed control. These make use of **a shunt-field rheostat, resistance in the armature circuit,** or **armature-voltage control.**

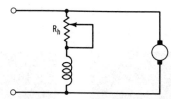

Fig. 3-21. Speed control using a shunt-field rheostat.

Look at Fig. 3-21. As the rheostat (R_h) is changed to increase the resistance in the circuit, **less current** flows in the field coil, and the speed of the motor **increases.** When the resistance is completely short-circuited, the full voltage is impressed on the field coil, and the motor rotates at its lowest possible speed.

If the resistance of the rheostat is increased, the strength of the main field will decrease. The counter emf generated in the armature will also decrease. The current through the armature will then increase, and the speed of the motor will increase. When the resistance is decreased, the opposite action takes place, and the motor slows down.

As a motor speeds up, the commutator begins to spark (every commutator has a practical speed limit), and the mechanical stresses within the rotor increase sharply due to centrifugal force and vibration. If the speed continues to increase, the armature may fly apart.

Armature-resistance speed control is obtained by inserting an external variable resistance in series with the armature circuit, as shown above. Speed control for series motors is usually not required since they are used where varying speed is per-

Q3-34. Two types of time-delay relays depend on a _ _ _ _ _ _ _ _ _ _ _ _ or a _ _ _ _ _ _ _ to determine the amount of time delay.

Q3-35. If a motor requires an input of 30 amperes at 230 volts when its output is 8 hp, what is its efficiency?

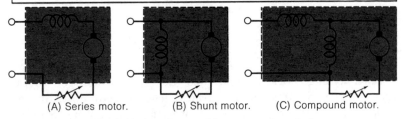

(A) Series motor. (B) Shunt motor. (C) Compound motor.

Fig. 3-22. Armature-resistance speed control.

missible. While this system meets the requirement of excellent speed control, the main disadvantage is the considerable amount of power loss in the control rheostat. This, in turn, makes the motor efficiency very low.

Ward Leonard Speed Control

The speed of a dc motor can be controlled by varying the armature voltage. This is done by the Ward Leonard system shown in Fig. 3-23. This system is the most accurate but also the most expensive method of motor-speed control.

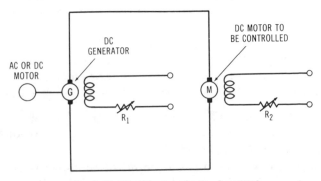

Fig. 3-23. Ward Leonard speed control.

The shunt fields of both the generator and motor are supplied from a constant-voltage source. The dc-generator output voltage is controlled by rheostat R_1. The voltage variation (out of G and into M) gives a wide range of speed control. In addition, by varying the motor rheostat (R_2), a wider speed range can be obtained.

This system requires three machines instead of one. A source of voltage for the field windings is also needed. In some cases, this may be a fourth machine. However, in many cases, the higher cost of this system is justified by the complete range of speed control it makes possible. With it, the speed can be varied accurately from zero to maximum.

ELECTRONIC SPEED CONTROLS

Dc motors can be controlled very effectively by using controllers that employ electronic devices such as transistors and thyristors. These electronic devices were described in earlier volumes of this series. With electronic control, a smoother and more efficient control of dc and ac motors is possible, and they can be controlled better than with any other system.

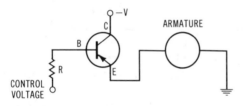

Fig. 3-24. A speed-control circuit for a small motor.

Fig. 3-24 illustrates one basic method of speed control for a small motor. A pnp transistor is used in an emitter-follower configuration. If a voltage applied to the base (B) through re-

Q3-36. What are the mechanical losses in a dc motor?

Q3-37. Why is more power lost in an armature speed-control rheostat than in a shunt-field rheostat?

Q3-38. When the shunt-field resistance is increased, the speed of a motor _ _ _ _ _ _ _ _ _ .

Q3-39. When the armature resistance is increased, the speed of a motor _ _ _ _ _ _ _ _ _ .

sistor R is varied, the current through the transistor collector-emitter junction will change in proportion, and so will the current through the motor that is in series with the transistor. The change in current varies the speed of the motor in proportion to the voltage applied to the base terminal of the transistor.

The basic circuit of a second electronic motor-control method is shown in Fig. 3-25. It uses an SCR for controlling the amount of current passing through the motor and, thus, controlling the motor speed. The SCR control circuit conducts only during a part of the positive half-wave portion of the ac input voltage. A pulse of voltage, which occurs at a 60-Hz rate, is applied to the gate of the SCR. Notice that the pulse does not necessarily turn the SCR on immediately as each positive half cycle occurs. The timing of the gate pulse determines when turn-on occurs during each half cycle. The sooner the gate pulse occurs, the greater the average current through the SCR, and, thus, the greater the speed of the motor.

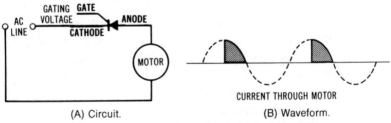

(A) Circuit.   (B) Waveform.

Fig. 3-25. A speed-control circuit that uses an SCR.

Another SCR motor-control circuit, shown in Fig. 3-26, uses two diodes (CR1 and CR2) to provide full-wave rectification of the ac line voltage. The timing of each pulse going to SCR1 and SCR2 determines at which point on the positive half wave the corresponding SCR will conduct. Thus, the timing of the pulses control the speed of the dc motor. Electron current flows through the motor from point 1 to point 2, whether it is SCR1

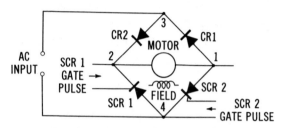

Fig. 3-26. An SCR motor-control circuit.

or SCR2 that is conducting. When the ac line voltage is positive at point 3, electrons flow from point 4 through the base to the anode of SCR2 (assuming SCR2 is gated "on"), and then to point 1, through the motor armature to point 2, through diode CR2 to point 3.

When point 3 in the circuit is made negative, electrons will flow from point 3 through diode CR1 to point 1, then through the armature to point 2, and through SCR1 to point 4, which is the positive side of the ac line at that instant. It should be noted that, for each condition, each SCR must be gated on by its gating pulse.

One main advantage of an electronic control is that no rheostat is used, thus eliminating a major cause of power loss in the controlling element.

Q3-40. The Ward Leonard system controls the motor speed by varying the _ _ _ _ _ _ _ _ _ _ _ _ _ _ _.

Q3-41. Which type of speed control gives full-range control of speed?

Q3-42. For an SCR to control the speed of a motor, a _ _ _ _ _ _ pulse must be put on the gate of the SCR.

Q3-43. The _ _ _ _ _ _ of the gating pulse determines at which point on the ac wave the SCR will turn on or "fire."

BRUSHLESS DC MOTORS

Numerous dc motors are found in consumer and industrial use. This is especially true where variable speed, reversing, positioning (as in servomotors), or battery operation is needed.

A disadvantage of dc motors is the need for frequent maintenance due to brush wear. Brushes must be replaced periodically, and they cause dust which deteriorates the bearings. Also, arcing between brush and commutator causes electromagnetic and radio-frequency interference. Further, the rotational speed of a dc motor is limited due to the mechanical contact bounce of the motor brushes as the commutator is turning.

For these reasons, a special kind of motor, called a brushless dc motor, is now being used in many low-power applications. There are several types, but none use a commutator-brush arrangement. Some use electronic means (transistors or SCRs) to switch the field as the rotor turns. Others use a photosenser and photoemitter to sense the rotor position and switch the field at the proper instant. A few use an equivalent magnetic method, while others use a rotating magnet and reed switches combination.

The circuit of a typical brushless dc motor system is shown in Fig. 3-27. Although, as a system, operation is from a dc source, the motor used is a single-phase, permanent-magnet, split-phase induction motor operated by an oscillator circuit that is formed around transistors Q1 and Q2. The main winding of the motor is center tapped; transistor Q1 conducts

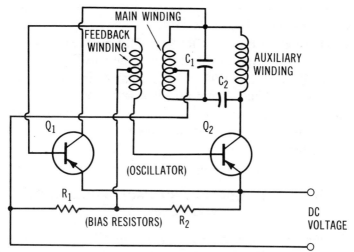

Fig. 3-27. A brushless dc motor circuit.

through one leg of the main winding and, then, transistor Q2 conducts through the other leg, and so on, alternately. Capacitor C1 is placed across the main winding to tune with the main winding as a resonant circuit, thus, holding the motor at its operating frequency. Capacitor C1 also absorbs voltage peaks and transients. Resistors R1 and R2 bias the transistors. Capacitors C1 and C2 make up the split-capacitor feature of the motor. A feedback winding in the stator slots provides a control voltage and determines the oscillator frequency. Both C1 and C2 are external capacitors and can be changed to vary the speed of the motor. For precise speed, either a tuning fork or a crystal and countdown circuit can be used. Brushless dc motors have none of the shortcomings previously mentioned for regular dc motors, and they can operate at speeds up to 10,000 rpm or higher.

Q3-44. Diode rectifiers CR1 and CR2 provide _ _ _ _-_ _ _ _ rectification in the SCR control circuit shown in Fig. 3-26.

Q3-45. Most of the problems of regular dc motors are due to the _ _ _ _ _-_ _ _ _ _ _ _ _ _ _ arrangement.

Q3-46. A special type of dc motor called a _ _ _ _ _ _ _ _ _ motor eliminates the disadvantages of regular dc motors.

ELECTROMAGNETIC AND RADIO-FREQUENCY INTERFERENCE

The increased use in recent years of the electromagnetic spectrum, from low frequencies up through microwaves and beyond, has greatly heightened the importance of preventing any generation of harmful interference. Interference is genrally of two types: electromagnetic interference (EMI) and radio-frequency interference (RFI). Keeping both types of interference as low as possible is not only important in the reception of radio-television signals in the home, but it is especially important in the aeronautical and aerospace fields, the government and military services, and the medical and scientific fields.

Interference is transmitted basically by two means. These are power-line conduction and electromagnetic radiation. In power-line conduction, the interference from the originating equipment travels through its connection to the power line into the second equipment, thus causing interference to the circuits and operation of that equipment. In the second means of transmission, the undesired signal is radiated into surrounding space by the interfering equipment and the generated signal is picked up on the receiving antenna or input circuits of the second piece of equipment.

The importance of keeping interference at a low level has resulted in a greater attention in the United States to the interference problem and has resulted in the monitoring of interference by the Federal Communications Commission (FCC). The FCC has two sets of standards for EMI and RFI. One concerns residential or consumer equipment and one is interested in industrial equipment. We won't go into the details of the requirements and standards here. You can inquire of the FCC, Washington, DC, if you need this information. You should know, however, some of the basic methods of preventing or reducing interference because one of the greatest producers of interference is the brush-commutator action of dc motors. Other interference producers are SCR controls and digital computer equipment. These types of equipment include microprocessors and those electronic calculators that generate pulses.

Some basic methods of eliminating or reducing EMI and RFI are: (1) by the enclosure, shielding, and grounding of the circuits or equipment that causes the interference and, (2) by including filters in those circuits which pass signals into, within, and out of the equipment that generates the interference. In some cases, ferrite beads are also used around supply leads to absorb RFI.

Fig. 3-28 illustrates how an SCR control circuit may be filtered and shielded against interference. Capacitors C1 and C2

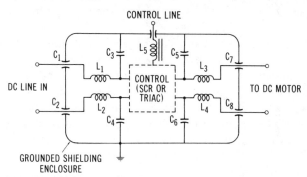

Fig. 3-28. One method for shielding and filtering an SCR control circuit.

are coaxial feedthrough capacitors used to prevent interference from entering the power source. Capacitors C7 and C8 are used to filter any interference from the load. Additionally,

Q3-47. The SCR-controlled brush-type dc motor is a serious producer of _ _ _ _ _ _ _ _ _ _ _ _ _ _ _ _ and _ _ _ _ _ - _ _ _ _ _ _ _ _ _ interference.

Q3-48. The basic methods of reducing EMI and RFI are _ _ _ _ _ _ _ _ _ , _ _ _ _ _ _ _ _ _ , _ _ _ _ _ _ _ _ , and _ _ _ _ _ _ _ _ _ .

inductors L1 and L2 and capacitors C1, C2, C3, and C4 provide a Pi-section filtering action within the enclosure at the input. Then, inductors L3 and L4 and capacitors C5, C6, C7, and C8 do the same at the output. Capacitor C9 and inductor L5 filter the control line. The SCR (or Triac) section is enclosed within a box that is additionally shielded.

WHAT YOU HAVE LEARNED

1. The main parts of a dc motor are the frame, end bells, pole pieces, shaft, armature assembly, commutator assembly, and brush assembly.
2. A dc motor has losses like those of a dc generator: copper losses, eddy currents, and hysteresis losses.
3. Armature reaction occurs in a dc motor and can be limited by interpoles and compensating windings.
4. A motor, when it is turning, also acts as a generator and creates a counter emf that opposes the applied voltage and limits the flow of current through the armature.
5. There are three main types of dc motors—shunt, series, and compound.
6. Shunt motors have a low starting torque and a relatively constant speed regardless of load.
7. Shunt motors must not be operated without field current because the motor will then run at a dangerous speed.
8. Series motors have a high starting torque, but their speed varies greatly as the load changes.
9. Series motors must never be operated without a load because, under that condition, the speed will be so great that the armature may fly apart.
10. Dc motors are started with a resistance in series with the armature in order to limit the starting current.

11. Compound motors combine the characteristics of both series and shunt motors in proportions that depend on the construction of the particular motor.
12. Two common ways of providing starting resistance for dc motors are the three-point and four-point starting boxes. In these devices, a contact is moved by hand to bypass the starting resistance as the motor gains speed. A magnet then holds the contact in place while the motor is operating.
13. The faceplace starter, a common form of manual starter, is used for motors up to 50 hp.
14. A motor starter that includes both reversing and speed control is usually called a motor controller.
15. Reversing the rotation of a series motor can be done by reversing either the field or the armature connections.
16. A shunt motor is usually reversed by reversing the armature.
17. For motors over 50 hp, drum controllers are used.
18. Automatic starting boxes perform the starting function automatically by using relays to bypass the resistance as speed builds up or after a time delay.
19. Thermal overload relays are operated by the heating effect of a current and can be used to protect motors from overheating.
20. Dashpot time-delay relays use an oil-filled piston to slow down the action of the relay.
21. An RC circuit can be used to provide a time delay for the operation of a relay.
22. Motor efficiency is found by dividing the output power by the input power. One horsepower is equal to 746 watts.
23. Motor speed can be controlled by the use of a shunt-field rheostat.
24. Motor speed can also be controlled through the use of an armature speed-control rheostat. The disadvantage of this system is that it reduces motor efficiency considerably through power loss in the rheostat.
25. The most useful system of speed control is the armature-terminal–voltage speed control (the Ward Leonard system), which uses an independent generator to provide voltage to the armature. This system allows a wide range

of accurately controlled speeds, but it is complex and expensive.

26. Electronic controls provide a smoother and, because no rheostat is used, a more efficient operation of motors.

27. A brushless dc motor eliminates the brush-commutator problem that is commonly associated with dc motors.

28. Interference can be reduced by shielding, enclosing, grounding, and filtering the circuit or component that generates the interference.

4

AC Generators

The principles of ac generator operation are presented in this chapter. You will learn the applications of the various generator types and how to recognize them. You will become familiar with the characteristics of ac generators and how they are regulated.

ALTERNATORS

Ac generators are also known as **alternators**. They vary in size from no larger than a walnut to bigger than a house. Almost all electrical power for homes and industry is supplied by alternators in power plants. An ac generating system consists of the **armature, field,** and **prime mover.**

The Armature

The armature is that part of a generator in which the output voltage is induced. The current that flows to the load also flows through the armature. In an alternator, the armature is an assembly of coils, as in a dc machine. The armature may be either the rotating (rotor) or stationary (stator) member of an ac generator.

The Field

Direct current is supplied to the field winding in an alternator. This dc current creates a magnetic field which is cut by the armature winding. A separate dc source is usually used for the field. The field, in alternators supplying up to 50 KW, is usually the stationary part (stator). In larger machines, the field is the rotating component (rotor). Electrically, it makes no difference whether a rotating winding cuts a stationary field or a rotating field cuts a stationary winding.

The field requires relatively low voltage and current compared to the high voltage and current that is generated in the armature of a large alternator. It is easier to connect this low voltage and current to a rotor (through slip rings) than it is to connect a high ac voltage and current. This is why the field is the rotating part in large alternators.

The Prime Mover

The prime mover is the source of mechanical power that drives the rotor of the alternator. It can be a gasoline engine, a steam turbine, a water turbine, or any such source. The prime mover can even be an electric motor.

SYNCHRONOUS ALTERNATORS

The synchronous alternator is the basic and most common ac generator. The dc excitation is provided from an outside source, usually a small dc generator. The shaft of the alternator is driven at a constant speed—usually 1800 or 3600 rpm. As has been mentioned before, either the armature or the field of the alternator can be the rotor.

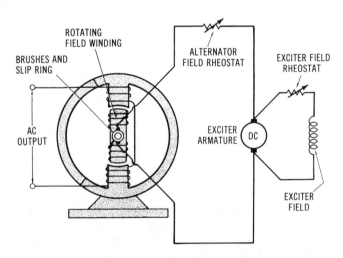

Fig. 4-1. Basic synchronous alternator.

Field Current

Dc current can be supplied to the field in several ways. In most cases, it is supplied by a dc generator called an **exciter**. The exciter may supply a dc line (called a **bus**) that is tapped to supply several loads. In other cases, the dc generator may be connected directly to the same shaft as the alternator, or it may be driven from the alternator shaft by a belt connection. Belt-connected exciters are used with relatively slow machines.

Since generators cannot operate without dc excitation, power plants usually have one or two spare exciters that are capable of taking over in case of an exciter failure. The exciter is usually a compound-wound dc generator that is flat compounded and is rated at either 125 or 250 volts, according to the size of the alternator. A flat-compound dc generator is used for excitation current because this type of machine gives the most constant voltage, regardless of load.

Inductor Alternator

Inductor alternators are used to produce voltages at frequencies between 500 and 10,000 hertz (cycles per second). They are used for supplying power to induction furnaces used for melting or heating metals.

In an inductor alternator, both the armature and the field are stationary. The only rotating element is a toothed steel rotor which distorts the magnetic field of both the field and the armature. The motion of the steel teeth produces a rapidly vibrating magnetic field which induces a very high-frequency voltage in the armature winding.

Q4-1. What type of alternator is used to supply household current?

Q4-2. What type of alternator would be used to obtain a 5000-Hz ac current?

Q4-3. What is the minimum number of generators that you could expect to find at an ac power-supply installation?

Q4-4. Why do large synchronous alternators have rotating fields and stationary armatures?

Q4-5. Why is a flat-compound generator used as an exciter for a synchronous alternator?

THE INDUCTION GENERATOR

The induction generator can be used to develop special frequencies for special applications. For example, some high-speed tools are operated by ac voltages at frequencies of 90, 100, or 180 Hz. Induction generators also supply ac voltages at 25 or 50 Hz. An induction generator may be considered as a device for changing the frequency of an alternating current since it uses a three-phase ac power source.

The power source is used to drive an ac motor. At the same time, it is used to establish a rotating magnetic field. In Fig. 4-2, the magnetic field itself rotates and the field coils remain stationary. (You will see how this is done in the following chapter when induction motors are discussed.) If the armature of the induction generator is standing still, the voltage induced in it will have the same frequency as the rotating magnetic field. The action is very similar to that of a transformer. However, if the armature is made to turn, the frequency of the induced voltage is no longer the same as the frequency of the field. If the armature turns in the same direction as the rotating field, the generated frequency will be lower. In fact, if the armature turns at the same speed as the rotating field, the induced frequency will be zero. When the armature is moving in a direction opposite from that of the rotating field, the relative speed is increased and the frequency is higher.

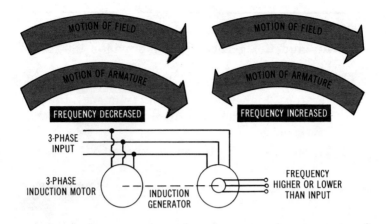

Fig. 4-2. Principle of frequency changing.

SINGLE PHASE AND POLYPHASE

Single phase refers to the type of generator discussed in Chapter 1. Single-phase alternators have only one armature winding whose leads deliver the output of the alternator. They deliver an ac output between two wires only.

Polyphase machines have more than one winding and deliver an output between several pairs of wires. The most common type of polyphase machine is the three-phase type. Two-phase machines are also sometimes used. Three-phase generators are widely used because large amounts of power can be transmitted more efficiently with a three-phase system than with a single-phase system.

Q4-6. What type of generator could be used to obtain a 120-Hz output?

Q4-7. When an induction generator supplies an output frequency higher than the input frequency, is the rotor turning in the same direction as the magnetic field?

Q4-8. What is the advantage of a three-phase power system over a single-phase power system?

Q4-9. A single-phase generator delivers voltage between only _ _ _ wires.

Two-Phase Generator

A two-phase generator is, as the name implies, a machine that has two separated windings on its armature. The two windings are usually mounted 90 electrical degrees apart. Thus, when the voltage in one coil reaches its peak, the other one is at zero, and vice versa. Fig. 4-4 shows the phase relationships in a two-phase generator.

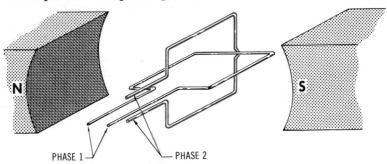

PHASE 1 — PHASE 2

Fig. 4-3. Simple two-phase generator.

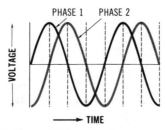

Fig. 4-4. Phase relations in a two-phase generator.

The armature windings of a two-phase generator can be connected in two different ways. It is possible to simply make individual connections to the two separate windings. This arrangement gives four output connections and is called a **four-wire system.** It is also possible, however, to combine two of the

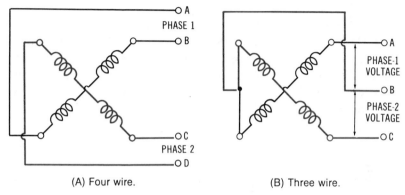

(A) Four wire.   (B) Three wire.

Fig. 4-5. Two-phase systems.

output connections to produce what is called the **three-wire system.** In this system, the voltages of the two separate phases remain the same. Phase 1 appears between points A and B. Phase 2 appears between points B and C. But a third voltage is also available. This voltage is the voltage between points A and C and is the **vector sum** of the other two voltages.

Fig. 4-6. Voltage relationships in a two-phase generator.

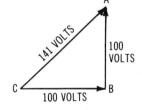

In Fig. 4-6, single-phase voltages of 100 volts exist between points A and B and between points B and C. The voltage between points A and C is 1.41 times the single-phase voltage if the two phase voltages are equal.

Q4-10. In a three-wire two-phase system, how can the combined voltage of the two phases be calculated if the individual phase voltages are equal?

Three-Phase Alternator

A three-phase alternator has three separate armature windings connected in either a **delta** or **wye** (Y-shaped) pattern. A detailed discussion of these methods of connection is given in Chapter 6 of this volume. (In Fig. 4-7, each phase winding is made up of two parts.)

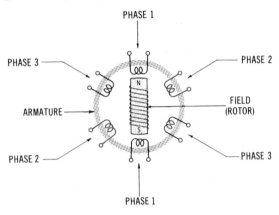

Fig. 4-7. Three-phase alternator.

The three single-phase voltages of a three-phase alternator are usually 120 electrical degrees apart. As in the two-phase machine, it is possible to obtain single-phase voltages and, also,

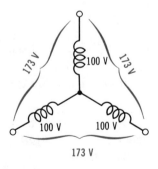

Fig. 4-8. Voltages in a wye-connected three-phase alternator.

voltages between phases. The relationship between these voltages depends on whether a delta or wye connection is used.

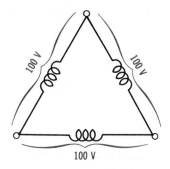

Fig. 4-9. Voltages in a delta-connected three-phase alternator.

100 V

As you will see in Chapter 6, three-phase power can be transmitted more economically than can single-phase power.

GENERATOR RATINGS

The ratings of an ac generator are usually given on a nameplate attached to the outside of the frame of the machine. The following information is usually given:

1. Manufacturer's serial and type numbers.
2. Number of phases.
3. Speed (rpm).
4. Number of poles.
5. Maximum voltage ratings (output).
6. Frequency of output.
7. KVA (or KW) rating.
8. Armature current (in amperes per phase).
9. Ambient temperature and temperature rise.
10. Field current (in dc amperes).
11. Power-factor limits.

Additional information, such as whether the generator can be operated continuously, may be given.

Q4-11. If the single-phase voltage of a three-phase wye-connected alternator is 120 volts, what is the voltage between phases?

Q4-12. How many between-phase outputs can be obtained from a three-phase machine?

ARMATURE REACTION

Armature reaction occurs in an alternator just as it does in a dc generator. The magnetic field of the armature interacts with the main field. With a purely resistive load, this effect is similar to that in a dc machine. There is some distortion of the main magnetic field which changes the waveform of the output voltage.

However, current in an ac circuit is not always in phase with the applied voltage. This means, also, that current in the armature is not always in phase with the induced voltage. This fact gives rise to an interesting effect. When the load is highly inductive and the current lags behind the load voltage, the phase shift of the current causes the magnetic field of the armature to oppose the main field and partly cancel it. If an inductive load is applied to an alternator, there is a drop in output voltage. Exactly the opposite happens with a capacitive load. Here, the load current leads the load voltage, and the armature field now adds to the main field. The output voltage of an alternator will be higher with a capacitive load.

FREQUENCY

In all alternators, the frequency is controlled by the speed of the rotor. The relationship between the speed of the rotor and the output frequency of an alternator is given by the following formula:

$$f = \frac{P \times S}{120}$$

where,
   f is the frequency in Hz,
   P is the number of poles,
   S is the speed in rpm.

This formula applies regardless of the number of phases.

FREQUENCY CONTROL

Generators must maintain very steady frequencies since so many electrical devices require an accurate supply frequency. For example, all electric clocks depend on an accurate frequency in order to maintain the correct time. A variation of only 1 cycle per second would mean a change of 24 minutes every 24 hours. Many devices, such as timers, are operated by synchronous motors because of the constant speed of this type of motor. The constant speed of a synchronous motor depends directly on a constant-frequency input.

The frequency of an alternator depends on the speed of the prime mover. If the steam turbine, hydraulic turbine, or fuel engine driving the generator has a reliable speed regulation, the generator frequency will be constant.

In large power plants, very accurate frequency-recording instruments and the means of compensating for any speed changes are maintained. Thus, if the frequency should drop for a short period, a control device will overspeed the shaft to make up for the loss. Fig. 4-10 shows a frequency-time diagram for such automatic correction. At time t_0, a situation

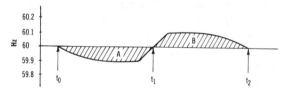

Fig. 4-10. Results of automatic frequency control.

such as a heavy load lowered the frequency to 59.9 Hz. At time t_1, the automatic device sensed the loss and forced the frequency to 60.1 Hz until time t_2. At that time, the machine had caught up with itself. Shaded areas A and B (on the diagram) must be equal in order for the machine to be caught up.

Q4-13. What would be the frequency of a 4-pole alternator operated at 1500 revolutions per minute?

Q4-14. Does the number of phases affect the frequency of the output? Why?

Q4-15. What happens to the output voltage of an alternator when the load is capacitive?

VOLTAGE REGULATION

The best way to understand the need for regulating the output voltage of an alternator is to determine what would happen if the voltage were not steady. One of the most disturbing things would be the constant flickering of electric lights. Certain motors would not maintain a constant speed. Radio and tv sets would not operate properly. The voltage of an alternator depends on the speed of the machine, the number of turns in the winding, and the strength of the magnetic field.

The speed of the shaft is maintained constant in order to maintain a constant frequency, so speed variation cannot be used to regulate voltage. The number of turns on the armature is fixed by the machine design and cannot be varied to regulate voltage. The field strength is the only other factor that can be varied, but even it can be varied only a limited amount. The graph in Fig. 4-11 shows how the no-load voltage of a typical alternator depends on the dc field current. An alternator operating at no load requires a minimum amount of field current—say 12 amps dc (point A in Fig. 4-11). If the field current is increased to 14 amps, the voltage will be increased to 135 V (point B on the graph).

Suppose the alternator is operating at 120 V with no load. A load is now applied to the generator. The voltage output will change if the field current remains the same. In this case, it drops to 110 volts (point C on the graph). Now, if the field current is increased to approximately 14 amps, the voltage will go up again to 120 volts (point D).

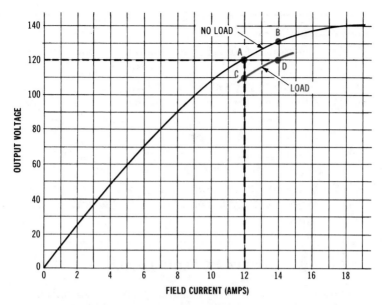

Fig. 4-11. Relationship of output voltage and field current in a typical alternator.

You can see that whenever a load is applied to an alternator, an increased field current should also be applied in order to make up for the drop in output voltage. During the course of one day, the load varies considerably. Such field-current changes must, therefore, be made automatically. A **voltage regulator** is a device used to regulate the output voltage of a generator.

Q4-16. Can the speed of an alternator be varied to control the output voltage? Why?

Q4-17. When a load is connected to an alternator, the output voltage of the alternator _ _ _ _ _ _ _ if the field current is not changed.

Q4-18. Changes in the output voltage of an alternator can be corrected by changing the _ _ _ _ _ _ _ _ _ _ _ _.

Q4-19. A device that automatically maintains a constant generator output voltage is called a _ _ _ _ _ _ _ _ _ _ _ _ _ _ _ _ _.

VOLTAGE REGULATORS

A voltage regulator must sense any change in output voltage and vary the dc field current so as to correct that change. There are many voltage regulators on the market but they can be divided into two basic groups—those with moving parts and those without moving parts. An example of the first type will be given later in this section. For simplicity, only the basic parts will be shown.

Fig. 4-12 shows the voltage sine wave during a voltage-regulating operation. The voltage begins to drop at time t_0. The voltage regulator senses the drop and at time t_1 begins to operate. The voltage first goes up and then gradually returns to

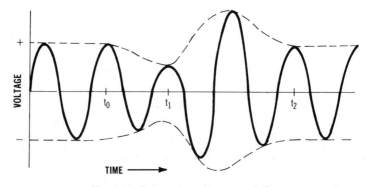

Fig. 4-12. Automatic voltage regulation.

normal. The time required for the voltage to return to normal varies from a few cycles, as shown in Fig. 4-12, to a few seconds.

An example of a mechanical voltage regulator is shown in Fig. 4-13. If the voltage rises, the magnetic field of the coil increases and the steel piece moves toward the coil. This causes some of the spring contacts to open, and the resistance between

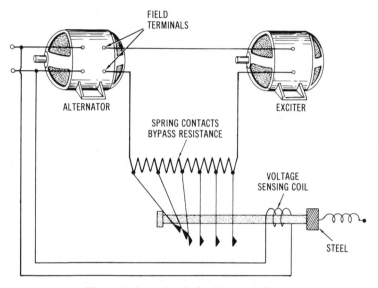

Fig. 4-13. A mechanical voltage regulator.

the exciter and the alternator increases. This causes less field current to flow to the alternator and the output voltage decreases to normal. If the voltage falls, the spring is able to pull the steel piece away from the coil. This closes more of the contacts. The resistance then decreases and the field current and output voltage increase. The resistance could also be placed in the exciter shunt-field circuit.

Q4-20. In the voltage regulator shown in Fig. 4-13, why is the resistance decreased when more of the contacts are closed?

Q4-21. Why does decreasing the resistance increase the output voltage?

PARALLEL OPERATION

Most power plants have several generators operated in parallel. The advantage of this method is that it provides more reliable operation.

In a large power distribution system, it is possible that power used in one area may come from generators operating several states away. This is because many power companies have their networks interconnected. In any large network, the line voltage is kept constant by the individual voltage regulation of each generator.

In order for generators to operate in parallel, their frequencies must be equal, their voltages must be equal, and they must be in phase with each other (synchronized). While most power plants are strictly three-phase systems, the following paragraphs will deal with a single-phase system for better understanding.

When a generator is being paralleled with others, it is said to be brought "on the line." It must first be brought to line voltage and proper frequency, and then it must be synchronized. This means that its voltage and the line voltage must go through the same parts of their cycles at the same time.

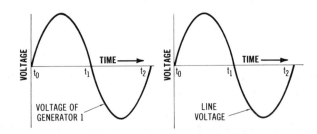

Fig. 4-14. Outputs of two synchronized generators.

There are many methods of synchronizing generators, but the most common is by the **lamps method.** (Other methods include the use of a **synchroscope** and the use of a **voltmeter** connected between the generator leads or across the synchronizing lamps.) A circuit showing the "lamps method" is given in Fig. 4-15. Generator G_2 is supplying the load, and G_1 is to be brought on the line.

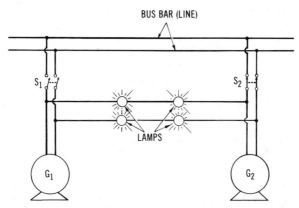

Fig. 4-15. The "lamps method" of synchronizing generators.

The voltage of generator G_1 is brought up to line voltage. If the bulbs are lighted, there is a voltage difference between the generators and, therefore, they are **not** in phase. In actual practice, the lights flash rapidly at first, as the operator adjusts the speed of G_1, then more and more slowly until they become dark. At this point, the operator closes switch S_1 and the machine is on the line. The lights flash because the voltages are going in and out of phase with each other when the generators do not have the exact same frequency. Considerable damage may result if the switch is closed when the lights are flashing, so this operation must be done only by qualified operators.

Q4-22. Why do the lights flash instead of remaining constantly lighted?

Q4-23. What three conditions must be met before two ac generators can be paralleled?

Q4-24. When synchronizing generators by the method shown in Fig. 4-15, what indicates that it is not safe to close the switch?

Paralleling Three-Phase Generators

The same general principles apply when paralleling three-phase generators as apply when paralleling single-phase generators. The voltages must be the same, the frequencies must be equal, and the generators must be synchronized. Care must

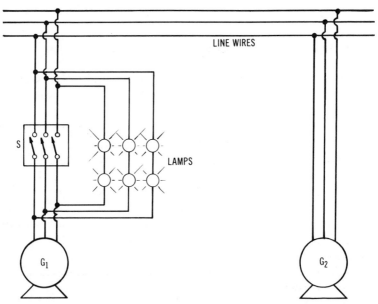

Fig. 4-16. Phasing out a generator.

be taken that the three phases are in the right sequence. If the phases are reversed in the generator being placed on the line, it will be damaged. In order to avoid this situation, the following test is performed before paralleling the machine. This is called **phasing out** the generator.

Look at the sketch in Fig. 4-16. With both generator G_1 and generator G_2 operating at the proper voltage and frequency, all lights must flash together and grow dark together. If they take turns in flashing, the connections to two of the phases of generator G_1 must be exchanged. When all the lights are dark, the machines are synchronized and switch (S) may be closed. This brings generator G_1 on the line.

When generators are operating in parallel, the frequency of all the generators remains the same. If one generator were to try to speed up or slow down, it would immediately be overloaded or underloaded and would be returned to the correct speed. For instance, say that generator G_2 in Fig. 4-17 tries

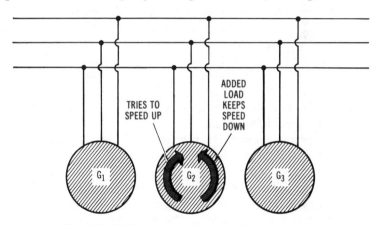

Fig. 4-17. All three generators have the same speed.

to speed up. An additional load is immediately placed on this generator. This additional load tends to hold the speed down. This is the action that keeps all of the generators operating at the same speed.

Q4-25. Why is it necessary to phase out a three-phase generator before paralleling it with another three-phase generator?

Q4-26. What is indicated if all the bulbs do not flash at the same time when a generator is being phased out?

WHAT YOU HAVE LEARNED

1. Ac generators are called alternators and are composed of three basic parts—armature, field, and prime mover.
2. The armature of an alternator is the assembly of coils in which the output voltage and current are induced.
3. The armature is often the nonrotating part of an alternator because it is easier to take high voltages and currents from it in this arrangement.
4. The field of an alternator is produced by a dc current in the field windings.
5. The prime mover of an alternator can be any sort of device that turns the generator shaft.
6. The synchronous alternator is the basic ac generator.
7. A synchronous alternator is usually supplied with dc exciter current from a flat-compounded dc generator which is often coupled to the shaft of the alternator.
8. The inductor alternator is used to generate very high frequencies by means of a vibrating magnetic field.
9. The induction alternator is used to convert normal ac frequencies to higher or lower frequencies for special purposes. It uses three-phase alternating current to set up a rotating magnetic field and a motor to vary the relative motion of the armature and the field.
10. A single-phase alternator delivers a single sine-wave output between two terminals.
11. Polyphase alternators have several independent windings and provide several sine-wave outputs.
12. The outputs of a two-phase alternator are normally 90 electrical degrees apart.

13. The two outputs of a two-phase alternator can be connected separately in a four-wire system, or one side of each output can be combined to provide a three-wire system.

14. In a three-wire two-phase system, the two equal voltages of each phase combine to give a between-phase voltage 1.41 times the single-phase voltage.

15. A three-phase alternator has three phases 120° apart.

16. The between-phase voltage of a three-phase wye-connected machine is 1.73 times the single-phase voltage.

17. Three-phase power is more efficiently and easily transmitted than single-phase power.

18. The type of load used will cause the voltage change that is due to armature reaction to vary. An inductive load causes the voltage to drop; a capacitive load causes the voltage to rise.

19. The frequency of an alternator depends on the number of field poles and the speed; it can be found by the formula, $f = \dfrac{P \times S}{120}$. P is the number of poles and S is the speed in rpm.

20. Large power plants have devices for automatic frequency control because the speed of most motors, clocks, etc., depends on frequency.

21. Power plants also have automatic voltage regulation to provide a relatively constant voltage output.

22. The output voltage of an alternator is regulated by varying the dc exciter current with a mechanical or electronic voltage-sensitive device.

23. To be placed in parallel, alternators must be at the same voltage, at the same frequency, and synchronized.

24. Lamps can be used to synchronize alternators. When the lamps connected between the lines to be paralleled all flash at the same time and, then, go dark, the two machines are synchronized. There is no voltage difference between the lines when the lamps are dark.

5

AC Motors

In this chapter, you will learn how a three-phase ac power supply is used to create a rotating magnetic field. You will discover how this rotating field is used to turn synchronous and induction motors. You will find out how to recognize these motors, and you will learn about their characteristics and applications. You will become familiar with the basic types of starting devices used with ac motors, when they are needed, and their advantages and disadvantages.

THREE-PHASE FIELDS

Remember that single-phase power is generated when a single constant magnetic field is rotated through a single winding (or vice versa). Three-phase power is generated in a similar manner when a magnetic field rotates in a three-phase winding. When single-phase power is fed into a single-phase winding, only a pulsating magnetic field is created. But, a rotating magnetic field is created when three-phase power is fed into a three-phase winding.

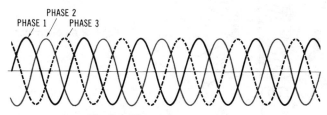

Fig. 5-1. Three-phase currents.

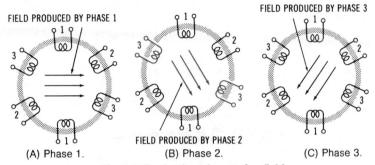

FIELD PRODUCED BY PHASE 1

FIELD PRODUCED BY PHASE 3

FIELD PRODUCED BY PHASE 2

(A) Phase 1. (B) Phase 2. (C) Phase 3.

Fig. 5-2. Production of a rotating field.

Fig. 5-2 shows the direction of the magnetic field that is produced by each phase winding. In Fig. 5-1, the current in phase 1 does not stop suddenly, nor does the current start suddenly in phase 2. In other words, the phase-2 current is increasing at the same time that the phase-1 current is decreasing. In the same way, the phase-1 field does not suddenly disappear and the phase-2 field does not suddenly appear. The phase-1 field decreases while the phase-2 field increases. The combined field does not jump suddenly from the position shown in Fig. 5-2A to the position shown in Fig. 5-2B. Instead, it moves smoothly from one position to the other as the phase-1 current decreases and the phase-2 current increases. In a similar way, the field moves from the position shown in Fig. 5-2B to the position shown in Fig. 5-2C as the phase-2 current decreases and the phase-3 current increases. The magnetic field produced by the three-phase winding rotates just as the field winding in the alternator (of Fig. 5-3) rotates.

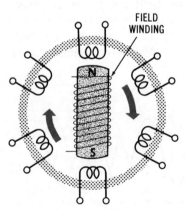

FIELD
WINDING

Fig. 5-3. An alternator.

SYNCHRONOUS MOTORS

Three-phase synchronous motors are similar in construction to three-phase synchronous generators. Both have a three-phase stator and a dc-powered field coil wound on the rotor. The electrical similarity can be seen from a comparison of the sketch in Fig. 5-4 with the drawing shown in Fig. 5-3.

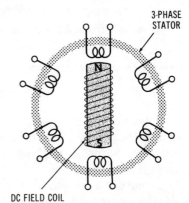

Fig. 5-4. Three-phase synchronous motor.

As the magnetic field in the stator rotates, the constant dc field rotates to keep aligned with it. As you can see, the speed of this motor depends on how fast the magnetic field rotates. The speed of the magnetic field depends on the frequency of the three-phase ac source. The synchronous motor cannot operate at any speed except that of the rotating field, which is called the **synchronous speed.** Synchronous motors are used where it is important to maintain a constant speed.

Q5-1. What sort of magnetic field is created in a single-phase winding fed with a single-phase ac current?

Q5-2. Three-phase power fed to a three-phase winding creates a _ _ _ _ _ _ _ _ magnetic field.

Q5-3. The speed of the rotating field produced by a three-phase winding is called _ _ _ _ _ _ _ _ _ _ _
_ _ _ _ _ .

Q5-4. What type of three-phase motor is used when it is necessary to maintain a constant speed?

POWER FACTOR

It was explained earlier that the power factor of the load determines how much dc field current an alternator will require to maintain a given output voltage. An inductive load (a lagging power factor) will cause a large voltage drop due to armature reaction and, therefore, a relatively high dc excitation is required in order to maintain a given voltage. A capacitive load (a leading power factor) will require a relatively low excitation because the armature reaction strengthens the main field. A resistive load will require a normal amount of excitation.

Something similar to what happens in an alternator happens in a synchronous motor. The power factor of a synchronous motor can be controlled by varying the field current. The amount of dc current required to cause a motor (at a given load) to operate at unity power factor is known as a normal excitation.

If the dc current to the machine is less than the normal value, the motor will operate with a lagging power factor (like an inductor). If the dc current to the machine is higher than the normal value, the motor will operate with a leading power factor (like a capacitor).

Synchronous motors are often operated at less than a rated load but with overexcitation so that they can help correct the power factor of a system. When a synchronous motor is operated at no load and is operated strictly for the purpose of correcting a system power factor, it is called a **synchronous capacitor.**

POLYPHASE INDUCTION MOTORS

The most important type of polyphase induction motor is the three-phase motor. Induction motors have two types of rotors, the **wound rotor** and the **squirrel-cage rotor**. The principle of operation is the same for both types.

A basic induction motor has neither slip rings nor a commutator. The rotating three-phase field of the stator induces a voltage in the rotor windings (hence, the name induction

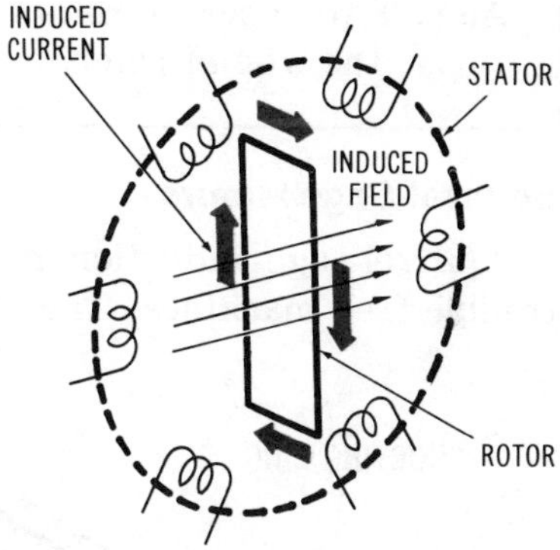

Fig. 5-5. Basic three-phase induction motor.

motor). This voltage, in turn, creates a large current in the rotor circuit. The current is large because the only resistance opposing it is the resistance of the wires. This high current in the rotor loop creates a magnetic field of its own. The rotor field and the stator field tend to attract each other. This situation creates a torque which spins the rotor in the same direction as the rotation of the magnetic field produced by the stator.

Q5-5. The power factor of a synchronous motor can be varied by varying the ____ ____ ____.

Q5-6. A synchronous motor operated without load and used only to provide power-factor correction is called a ____ ____.

Q5-7. Current is induced in the rotor of an induction motor by the rotating ____ ____ of the stator.

Q5-8. Two types of rotors used in induction motors are the ____ rotor and the ____-____ rotor.

Squirrel-Cage Motors

Squirrel-cage induction motors are extremely rugged and trouble-free machines. They have heavy copper bars around

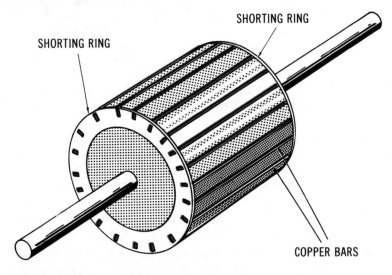

Fig. 5-6. A squirrel-cage rotor.

the rotor instead of a wire winding. The winding resembles a squirrel cage, from which it derives its name. The squirrel-cage motor has the advantages of requiring practically no maintenance and of costing much less than a wound-rotor motor.

Wound-Rotor Motors

In the wound-rotor motor, the rotor is not connected to any outside source of power. In order for this type of motor to operate, the terminals of all the rotor windings must be connected together, either directly or through a resistor bank. This is accomplished by means of brushes and by slip rings connected to the ends of each winding. In effect, this allows the operator to control the resistance of the rotor windings.

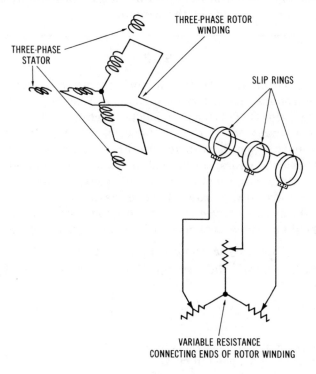

THREE-PHASE ROTOR WINDING

THREE-PHASE STATOR

SLIP RINGS

VARIABLE RESISTANCE
CONNECTING ENDS OF ROTOR WINDING

Fig. 5-7. A wound-rotor induction motor.

Q5-9. In a squirrel-cage rotor, _ _ _ _ _ _ _ _ _ _ serve as the conductors.

Q5-10. Name two advantages of squirrel-cage motors.

Q5-11. Do the brushes feed external power into the rotor of a wound-rotor motor?

SLIP

The speed of an induction motor can never be quite equal to synchronous speed. Synchronous speed is the speed of the rotating field (3600 rpm for 2-pole machines and 1800 rpm for 4-pole machines if the frequency is 60 Hz). If the rotor moved at this speed, no magnetic lines of force would move across its conductors and no voltage would be induced in the rotor. An induction motor cannot be operated at synchronous speed; it must always run a little slower.

The difference between the synchronous speed of the magnetic field and the actual speed of the motor is called **slip.**

$$\text{Slip} = \text{Synchronous speed} - \text{Actual speed}$$

Percentage of slip in an induction motor is given by the formula:

$$\% \text{ Slip} = \frac{S_S - S_A}{S_S} \times 100$$

where,

S_S is the synchronous speed in rpm,
S_A is the actual speed in rpm.

For example, what is the slip and the percentage of slip in a 2-pole, 60-Hz induction motor whose speed at full load is 3450 rpm?

$$\text{Slip} = \text{Synchronous speed} - \text{Actual speed}$$
$$= 3600 - 3450 = 150 \text{ rpm}$$

$$\% \text{ Slip} = \frac{3600 - 3450}{3600} \times 100$$
$$= \frac{150}{3600} \times 100 = 4.17\%$$

In most induction machines, the full-load slip varies from 4 to 6%. The number of phases, whether one, two, or three, does not matter when calculating slip values.

The resistance of the rotor circuit can be varied in a wound-rotor motor. The slip depends on this resistance (a greater resistance causes a greater slip). Therefore, it is possible to control the speed of a wound-rotor motor by choosing the proper resistance bank in the rotor circuit. When starting induction motors which have a load attached at all times (such as a flywheel), maximum starting torque can be provided by varying the resistance of the wound rotor to the correct value.

A **double-squirrel–cage** motor takes advantage of the same effect to obtain improved starting torque. This type of motor has two separate squirrel-cage windings. One has a high resistance to provide good starting torque; the other low.

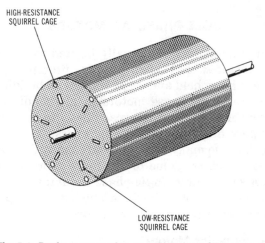

Fig. 5-8. Basic construction of a double squirrel-cage rotor.

Q5-12. What is synchronous speed?

Q5-13. What is slip?

Q5-14. If a 4-pole motor is supplied from a 60-Hz source and operates at 1750 rpm, what is the percentage of slip?

Q5-15. The speed and torque of an induction motor can be controlled by varying the _ _ _ _ _ _ _ _ _ _ of the rotor circuit.

SINGLE-PHASE AC MOTORS

Single-phase ac motors are usually limited in size to about two or three horsepower. There are many different types found in almost every household and every industrial building. In the home, there are single-phase motors in air conditioners, air heaters, refrigerators, sewing machines, fans, ventilating units, and many other household appliances.

Since they have so many uses, there are many different types of single-phase motors. Some of the more common types are **repulsion, universal,** and **single-phase induction** motors. Single-phase induction motors include **shaded-pole, split-phase, capacitor,** and **repulsion-induction** motors.

Single-Phase Induction Motors

Single-phase induction motors have no means of starting by themselves. In a single-phase motor, the field of the stator windings does not rotate as it does in a three-phase induction motor. The magnetic field set up in the stator by the ac power supply stays lined up in one direction. This magnetic field, though stationary, pulsates as the voltage sine wave does. The pulsating field induces a voltage in the rotor windings, but the rotor field can only line itself up with the stator field. Therefore, since the stator field is stationary, the rotor field is also stationary. Before it can start, the rotor must be turning so that there is, in effect, some slip and the two fields are not exactly lined up.

The single-phase induction motor acts much like the pedals of a bicycle (Fig. 5-9). When the pedals are exactly lined up with the direction of the up-and-down motion of the rider's feet, the pedals will not turn. Once a slight turn has started them, inertia carries the pedals past the center point, and the pulsating up-and-down motion keeps the rotating motion going.

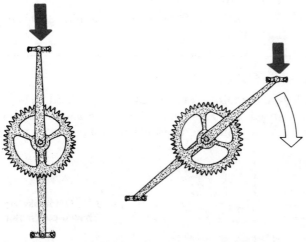

Fig. 5-9. Induction motor starting simulation.

It is necessary, then, to find some means of giving the single-phase ac motor a means of starting the rotor into motion. Once the rotor is spinning at a reasonable rpm, its inertia will carry it through the dead-center position so that it will be kept rotating by the stationary field. An **auxiliary starting system** is required. The starting methods make up the primary differences between induction-motor types.

Q5-16. **What is the largest size in which single-phase motors are usually made?**

Q5-17. **Name the three types of single-phase motors.**

Q5-18. **What are four types of single-phase induction motors?**

Q5-19. **The magnetic field of a single-phase motor (does, does not) rotate.**

Shaded-Pole Motors

Shaded-pole motors are usually very small. They have a constant speed and are usually in a size range of about 1/50 hp. The stator windings are arranged as shown in Fig. 5-10.

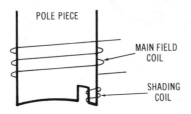

POLE PIECE

MAIN FIELD COIL

SHADING COIL

Fig. 5-10. Stator windings in a shaded-pole motor.

The shading coil is short-circuited. As the field in the pole piece builds up, a current is induced in the shading coil. This current causes a magnetic field that opposes the main field.

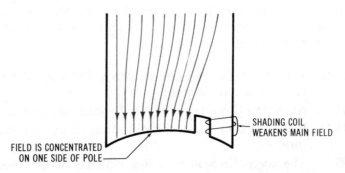

SHADING COIL WEAKENS MAIN FIELD

FIELD IS CONCENTRATED ON ONE SIDE OF POLE

Fig. 5-11. Action of shading coil as field strength increases.

The main field will, therefore, concentrate on the opposite side of the pole piece (Fig. 5-11). The result is that the field in the part of the pole piece inside the shading coil reaches a maximum intensity later than the rest of the field. As the field in the part of the pole opposite the shading coil begins to decrease, the shading-coil field will start aiding the main field and the concentration of flux moves to the other edge of the pole piece (Fig. 5-12). The resulting field, in the part of the pole piece inside the shading coil, reaches zero after the main field.

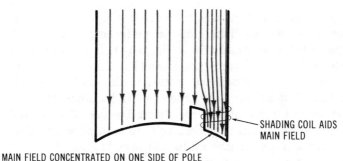

SHADING COIL AIDS
MAIN FIELD

MAIN FIELD CONCENTRATED ON ONE SIDE OF POLE

Fig. 5-12. Action of shading coil as field strength decreases.

The effect of the shading coil is to produce a small sweeping motion of the main field from one side of the pole piece to the other as the field pulsates. This slight rotating motion is enough to start the motor (Fig. 5-13).

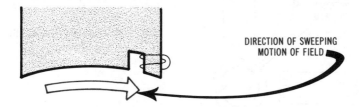

DIRECTION OF SWEEPING
MOTION OF FIELD

Fig. 5-13. Apparent motion of field produced by shading coil.

Q5-20. Shaded-pole motors usually have _ _ _ _ _ horse-power ratings.

Q5-21. The field, in the part of the pole piece inside the shading coil, reaches maximum _ _ _ _ _ the rest of the field.

Split-Phase Motors

Split-phase motors usually have ratings of ¾ hp or less. They have two separate coils—a main coil of large wire and a starting coil of small wire (Fig. 5-14). Both coils are placed in the motor in the same positions that they would have if the machine were a two-phase motor. If the coils have the same number of turns, they have the same inductance. However, the starting coil has a higher resistance because it is wound with smaller wire. When the same voltage is applied to both windings, the current in the main coil lags behind the current in the starting coil. The two windings produce a rotating field in much the same way that a rotating field is produced in a two-phase motor. This rotating field causes the motor to start.

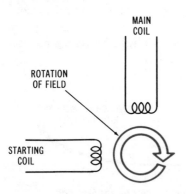

Fig. 5-14. Principle of the split-phase motor.

If used to run the motor, the starting winding is likely to burn out because of the small wire from which it is made. For this reason, a centrifugally operated switch is used to disconnect the starting winding when the motor reaches about 60% of its operating speed.

Capacitor Motors

Capacitor motors have two stator windings (Fig. 5-15). A large capacitor is connected in series with one of the windings. The single-phase input voltage produces two currents (one in each winding). One current is shifted out of phase with the applied voltage by the capacitor. This creates a starting torque similar to that of a two-phase motor.

Capacitor-Start, Capacitor-Run Motor—In this type of motor, the starting capacitor stays in the system at all times. This type of machine is made in sizes from ⅛ to 10 hp. It has a relatively high power factor.

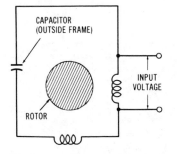 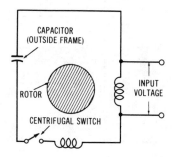

(A) Capacitor start, capacitor run. (B) Capacitor start, centrifugal switch.

Fig. 5-15. Capacitor motors.

Capacitor-Start Motor—This motor starts with a capacitor in series with one of the windings. At about 75% of full speed, a centrifugal switch opens the capacitor-winding circuit and the motor operates as a single-phase inductor motor. This type is made in sizes from ⅛ to ¾ hp.

Two-Value, Capacitor-Start, Capacitor-Run Motor—This type combines the features of both capacitor-start and capacitor-start, capacitor-run motors. It has two parallel-connected capacitors in series with one winding and a centrifugal switch which cuts out one capacitor at about 75% of full speed. This type of machine has ratings from ⅛ to 10 hp. It has many industrial applications.

Q5-22. **The various starting methods for single-phase induction motors all create a _ _ _ _ _ _ _ _ magnetic field.**

Repulsion Motors

The repulsion motor (Fig. 5-16) has an armature and commutator similar to that of a dc machine. The two brushes are
connected by a low-resistance wire.

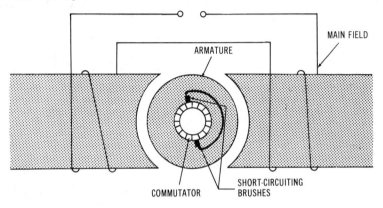

Fig. 5-16. Repulsion-induction motor.

The stator windings (actually two windings in series) produce a current in the rotor windings by induction. This current produces magnetic poles in the rotor, with their location
depending on the position of the brushes. When the field of the
rotor is at an angle with the main field, a torque is created by
the interacting fields. The rotor moves in a direction to turn
its field away from the main field, but as it moves, the brushes
come into contact with a different pair of commutator segments
and shift the field back.

There is no starting problem in a repulsion motor. They have
a good starting torque and are used where heavy starting loads
are expected.

Repulsion-Induction Motors

The repulsion-induction motor has a wound rotor with a
commutator. Shorting brushes make contact with the commu-

tator and the motor starts as a repulsion motor. As the motor nears full speed, a device short-circuits all of the commutator bars. The motor then runs as an induction motor and operates at nearly constant speed. Its speed cannot be adjusted. This type of motor is normally made in sizes ranging from ½ hp to 10 hp.

Universal Motors

One of the most versatile motors is the **universal motor** (Fig. 5-17). It operates on either dc or single-phase alternating current. These machines have high starting torque and high slip. They are usually in the fractional-horsepower range and are used in small appliances, electric drills, etc.

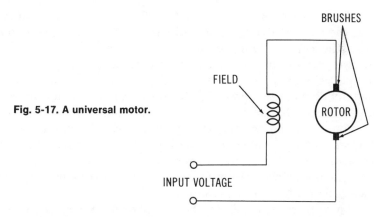

Fig. 5-17. A universal motor.

As you recall, a dc series motor will continue to turn in the same direction if the line connections are reversed. A series dc motor will therefore work on alternating current. In fact, a universal motor is simply a dc series motor whose windings and pole pieces are designed to operate efficiently with ac power.

Q5-23. Why is a centrifugal switch used in a split-phase motor?

Q5-24. A repulsion motor (does, does not) have a commutator.

Q5-25. Repulsion motors have a _ _ _ _ starting torque.

Q5-26. A universal motor can be operated on both _ _ and _ _ current.

INDUCTION-MOTOR STARTING

When voltage is first applied to a three-phase induction motor, the current drawn is sometimes six or seven times higher than the normal running current. However, this current decreases rapidly as the machine gathers speed. In the case of a squirrel-cage induction motor, the starting current will not usually damage the motor itself, but it may create an undesirable voltage fluctuation in the power system. It is therefore customary, when starting, to apply full-rated voltage only to the smaller squirrel-cage induction motors. Reduced starting voltage is applied to the larger-sized machines. The reduced voltage can be applied by means of an autotransformer, a series resistor, or a series reactor.

The autotransformer method is shown in Fig. 5-18. The group of autotransformers used to limit starting current is

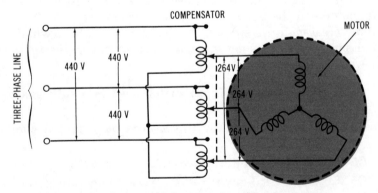

Fig. 5-18. Starting a squirrel-cage motor with a compensator.

called a **compensator.** When the motor reaches running speed, the compensator is bypassed and the full voltage is applied. The autotransformer will dissipate almost no power. (Only a negligible amount will be dissipated by the resistance of its windings.)

Another method of starting is to insert a resistor in series with each of the three motor windings, as shown in Fig. 5-19. The resistors will dissipate power.

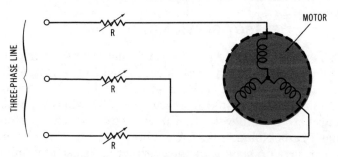

Fig. 5-19. Starting a squirrel-cage motor with resistors.

A third method of starting a squirrel-cage motor is to use series coils (reactors). A coil with a relatively high inductive reactance is inserted in series with each of the motor coils as shown in Fig. 5-20. When the motor has picked up speed, the switches are closed to bypass the reactors.

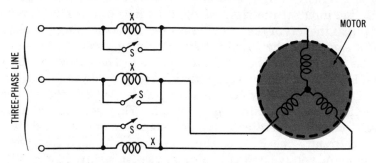

Fig. 5-20. Starting a squirrel-cage motor with inductors.

Q5-27. **Name three methods of limiting the starting current in an induction motor.**

Q5-28. **Would a ¼-horsepower, single-phase, squirrel-cage induction motor normally require a starter?**

ENERGY-SAVER MOTOR CONTROL

It has been estimated that electric motors use over one half of the electrical energy that is consumed in the United States on a day-to-day basis. Because of this fact and because of the need to conserve shrinking energy supplies wherever possible, researchers have developed various ways of reducing energy usage. Frank Nolan, a NASA engineer, has developed a circuit that can save up to 50% of the energy that is used by induction motors when operating at less than full load. One disadvantage of induction motors is that they use the same amount of current whether they are operating at full load or no load. Frank Nolan's circuit reduces the current used at no load, thus preventing much wasted energy.

The circuit shown in Fig. 5-21 is a simplified version of the actual NASA design. Power to the motor comes from the 60-Hz line input at the right of the drawing. Another 60-Hz line input at the left of the drawing is used for comparing the phase of a sample of the current that is passing through the motor with the phase of the actual line voltage. The output of the E vs. I comparator is a voltage that is fed to a power-factor control (PFC) circuit which then generates a voltage that is proportional to the desired phase angle. This voltage from the PFC circuit generates a trigger signal to the phase-control triac circuit. The PFC circuit senses the phase shift between the motor voltage and current and uses this shift to affect the "on" time of the triac. The "on" time of the triac, in turn, keeps the phase angle at the desired value, thus improving the power factor.

The circuit in Fig. 5-21 was first developed for small induction motors of 1 hp or less but modification work has been done on the circuit for industrial applications involving larger three-phase motors, such as industrial air-conditioning systems and pumps, large machine tools, and fans.

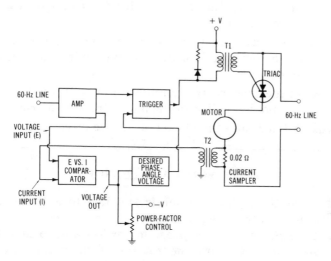

Fig. 5-21. Current reduction circuit.

Notice in Fig. 5-21 that the line voltage is applied to the motor through a triac. As mentioned in earlier chapters, a triac is a form of the thyristor. A triac can conduct current in either direction. A positive signal at the gate lets the triac conduct in one direction. A negative signal at the gate fires the triac in the other direction. Triacs are commonly used for light dimming, temperature control, and motor speed control. A triac can do the work of two SCRs.

Q5-29. The current required by a/an _ _ _ _ _ _ _ _ _ motor is approximately the same whether the motor operates at no load or full load.

Q5-30. The efficiency of an induction motor operating at no load can be improved by improving the _ _ _ _ _ angle.

Q5-31. In the energy-saver circuit of Fig. 5-21, the ac line voltage is applied to the motor through a _ _ _ _ _ _.

Q5-32. A triac can control current in (one, two) direction(s).

WHAT YOU HAVE LEARNED

1. When three-phase power is supplied to a three-phase winding, a rotating magnetic field is created.

2. A synchronous three-phase motor has a dc-excited field that interacts with the rotating field created by the three-phase ac power supply and it causes the rotor to turn at the same speed as the magnetic field.

3. Synchronous motors are used in order to provide constant speeds.

4. The power factor of a synchronous motor varies according to the amount of dc excitation. When the dc field current is below normal, the motor has a lagging power factor and behaves as an inductor. When the dc field current is at the normal excitation value, the machine presents a purely resistive load and has a power factor of one (unity). When the dc field current is above the normal excitation value, the motor has a leading power factor and acts as a capacitor.

5. In a polyphase induction motor, the rotating magnetic field induces a current in a short-circuited rotor winding, and motor action occurs when the field created in the rotor interacts with the main field.

6. There are two types of polyphase induction motors— squirrel cage and wound rotor.

7. A squirrel-cage motor has a rotor winding composed of copper bars embedded in the iron rotor core. A wound rotor has conventional wire windings.

8. Squirrel-cage motors are rugged and they require very little maintenance. Therefore, they are relatively inexpensive.

9. An induction motor cannot operate at synchronous speed. The difference between the speed of the rotating magnetic field and the motor speed is called slip.

10. The greater the resistance of the rotor winding of an induction motor, the greater is the slip.

11. Wound-rotor induction motors have a variable rotor resistance that is controlled by the operator and is connected to the rotor through an arrangement of slip rings and brushes.

12. Single-phase induction motors must have some arrangement to create a rotating magnetic field for starting the machine. Single-phase induction motors are classified according to the starting method used.

13. Shaded-pole motors create a rotating-field effect by means of a short-circuited shading coil on one edge of the pole piece. This coil produces a field that first weakens and, then, aids the main field.

14. Split-phase motors have a high-resistance starting winding whose current and field are more nearly in phase with the applied voltage than those of the main winding. The combined field of the two windings creates a rotating-field effect.

15. A centrifugal switch is used to disconnect the high-resistance starting winding after a split-phase motor has picked up speed. This is done so that the starting winding will not burn out.

16. Capacitor motors use capacitors in series with one of the windings to produce a phase shift similar to that of a split-phase motor.

17. Capacitor-start, capacitor-run motors keep the capacitive winding in the circuit at all times. Capacitor-start motors have a centrifugal switch that is used to disconnect the capacitive winding after the motor has been started.

18. Repulsion motors have a wound rotor and a commutator that is short-circuited by brushes. This motor has a high starting torque.

19. A universal motor is similar to a series dc motor.

20. Starting current for large induction motors is limited in one of three ways—with a compensator (autotransformers), with series resistors, or with series reactors.
21. Induction motors use just about as much power when operating at no load as at full load. Methods of improving the power factor at no load can increase the efficiency considerably.

6

Three-Phase Systems

what you will learn

In this chapter, you will learn how three-phase power is generated and distributed. You will become acquainted with the various ways in which three-phase alternators, motors, and transformers can be connected. You will be able to tell the difference between line voltage (and current) and phase voltage (and current), and you will learn how to find one if the other is known. You will also learn how to calculate and measure power in three-phase systems.

THREE-PHASE GENERATION AND DISTRIBUTION

Throughout the world, power plants produce huge quantities of electrical power in order to supply the ever increasing requirements for light, electric heating, and heavy industry. Nearly all of this power originates from three-phase generators. The voltage is stepped up by transformers for transmission and further transformed to lower and higher voltages according to need. The power is eventually used in either three-phase or single-phase devices.

Voltages and frequencies are usually standardized. In the United States, for example, the frequency is 60 hertz (cycles per second) for the major portion of the country. A frequency of 25 Hz was quite common at the beginning of this century, and a few 25-Hz units are still in operation. In Europe, both 50-Hz and 60-Hz frequencies are used, but the trend is toward adopting a uniform 60-Hz frequency.

The great majority of the alternators that produce three-phase power are synchronous generators. This type of generator is chosen because of its very high efficiency—as high as 99% for very large units—and for its ability to maintain a very steady voltage and frequency output.

All large power-generating systems have three basic components—a prime mover, a three-phase alternator, and a dc field exciter. The prime mover provides the mechanical power necessary to turn the rotor of the alternator and the exciter. Hydraulic turbines (driven by falling water), steam turbines, and diesel engines are all possible prime movers. Steam and hydraulic turbines are the most common.

The three-phase alternator produces three-phase electricity at a given voltage and frequency. The voltage output may be varied to some extent by changing the dc excitation value, and the frequency may be varied by changing the speed of the rotor. Voltage- and frequency-regulating equipment is used to keep these quantities constant.

The dc field excitation provides control of the output voltage. The direct current is monitored and controlled by a voltage regulator. The dc generator is usually mounted directly on the same shaft as the alternator or coupled to the alternator shaft by a belt drive.

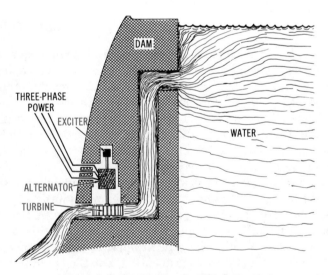

Fig. 6-1. Hydroelectric generating plant.

Water-driven generators are operated at low speeds. A typical speed is 200 rpm. Falling water enters the turbine case and spins the turbine which is connected to a shaft. This shaft rotates the alternator and exciter rotors. All units are usually mounted vertically. Such a generating system is called a **hydroelectric plant.**

The speed of a steam-driven generator is normally 3600 rpm (for 60 Hz). Some generators operate at 1800 rpm, although these are less common. In a steam-generator plant (Fig. 6-2), the components are arranged horizontally. Steam from an oil- or coal-fired boiler system, or from a system heated by a nuclear reactor, is piped to the turbine. The steam expands and pushes against the blades causing the turbine shaft to rotate. The shaft turns the rotors of the alternator and exciter.

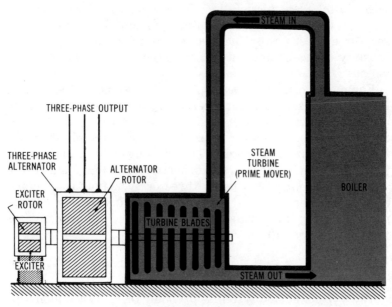

Fig. 6-2. Steam generator plant.

Q6-1. The standard frequency for alternating current in most of the world is ——.

Q6-2. The three essential parts of a generating system are the _ _ _ _ _ _ _ _ _ _, _ _ _ _ _ _ _ _ _ _, and _ _ _ _ _ _ _.

Transmission of Power

After power is generated at the power plant, the voltage is usually stepped up for transmission. Voltages as high as 69,000 volts are common. By stepping the voltages up to these high values, it is possible to transmit large amounts of power with relatively low currents. The low currents can be carried by smaller wires in the transmission lines. Less copper is needed and less power is lost. (Power loss in a line is equal to I^2R). As you will see, three-phase power lines use less copper than lines designed to carry the same amount of single-phase power at the same voltage.

For most household uses, the three-phase power is split up into single-phase, 120-volt alternating current. For many industrial applications, it is used as three-phase power. For example, three-phase power is used to drive three-phase induction and synchronous motors. The winding connections used in three-phase alternators, transformers, and motors make it possible to transmit power using three or even four conductors. Since the systems using these types of winding connections appear wherever three-phase power is used, it is important for you to understand them.

WYE CONNECTION

The **wye-connected** system is probably the most common type of three-phase connection. The wye connection is also called a **star connection.** It is usually diagrammed as shown in the sketches of Fig. 6-3.

Generators, motors, transformer windings, capacitors, resistors, etc., can all be connected in the same arrangement. Each phase is 120 electrical degrees away from the other two phases. The diagram resembles the letter Y, from which the name "wye" was taken.

The three windings in a wye connection are not only 120 electrical degrees apart, but they are also connected to a single common point. The ends of the three arms of the Y represent the three external ends of the windings. The center of the Y represents the three ends that are connected together and is called the **neutral point** or **common point.**

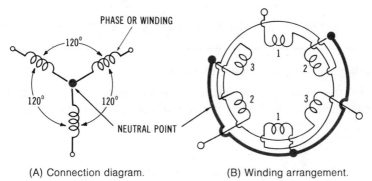

(A) Connection diagram.   (B) Winding arrangement.

Fig. 6-3. Wye connection.

The neutral point is normally grounded (connected to the earth or a large mass of metal). In this case, the wye wiring connection is as shown in Fig. 6-4. This system is called a three-phase four-wire wye connection.

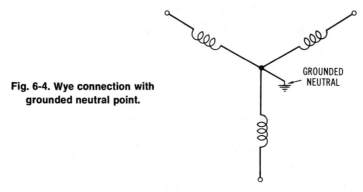

Fig. 6-4. Wye connection with grounded neutral point.

Q6-3. Why is the generator voltage stepped up before power is transmitted?

Q6-4. In a wye connection, one end of each phase or winding is connected to the _ _ _ _ _ _ _ point.

Q6-5. The neutral point is usually _ _ _ _ _ _ _ _ .

Voltage in a Wye-Connected System

In the case of a wye-connected generator with grounded neutral point, there are several voltages that can be measured and its outputs.

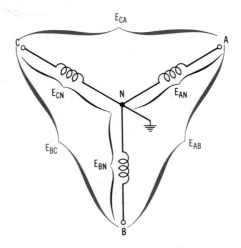

Fig. 6-5. Voltages in a wye-connected system.

These voltages are shown in the diagram of Fig. 6-5 and are as follows:

E_{AB}—The voltage between A & B.
E_{CA}—The voltage between C & A.
E_{BC}—The voltage between B & C.
E_{AN}—The voltage between A & neutral (N).
E_{BN}—The voltage between B & neutral.
E_{CN}—The voltage between C & neutral.

In most cases, the voltages between each phase terminal and the neutral point are equal. Thus, $E_{AN} = E_{BN} = E_{CN}$. The voltages between any two phases will then also be equal. Thus, $E_{AB} = E_{BC} = E_{CA}$.

The diagram in Fig. 6-6 indicates how the voltages add. Just as the distance from point A to B is greater than the distance from points A to N, the voltage between points A and B is greater than that between points A and N, and so on. The voltage between any one phase terminal and the neutral point (say E_{AN}) is considerably lower than any phase-to-phase voltage (say E_{AB}).

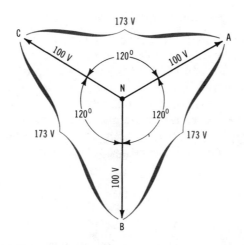

Fig. 6-6. Voltage relationships in a wye-connected system.

The phase-to-phase voltage is the between-the-lines voltage, or simply the **line voltage.** The voltage between a line and the neutral point is the **phase voltage.** Because of phase relationships, the line voltage is 1.73 times the phase voltage in a wye-connected system, if the phase voltages are equal.

Q6-6. The voltage between any two lines of a wye-connected three-phase transmissions system is called the ____ _____.

Q6-7. The voltage between one line and ground is called the _____ _____.

Q6-8. How are these voltages related?

Current in a Wye-Connected System

The current flowing in a given line is called the **line current.** The current flowing through any one winding is the current in a single phase and is called the **phase current.**

The same current that flows through a line also flows through the windings to which it is connected in a wye system. Thus, the line current is equal to the phase current in a wye system. This relationship applies to each phase individually. Current in one line does not necessarily equal current in another phase.

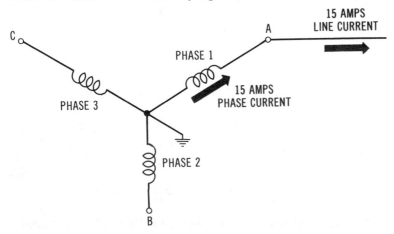

Fig. 6-7. Current relationships in a wye-connected system.

Summary of Wye-Connected Systems

If a three-phase wye-connected system is operating in the normal way, and if the phase voltage from A to N is 100 V rms, the voltages from B to N and C to N will also be 100 V rms. The line voltage from A to B will be 1.73 times the phase voltage, or 173 V rms. Line voltages from A to C and from B to C will also be 173 V rms. If the current through line A is 15 amperes, the phase current through winding A will also be 15 amperes.

Note that the **effective** values of voltage and current are usually used. These are the **rms values** (also called the root-mean-square values). The rms values determine how much power is dissipated. The same relationships between line and phase voltage and between line and phase current are also true if you wish to consider peak values. Remember that when comparing voltages or currents, they must all be in the same units (peak, rms, etc.).

Here is a typical problem. Suppose a resistive load, as shown in Fig. 6-8, is supplied power from a three-phase wye-connected alternator. The voltage between A and C is 208 V and the current in each line is 10 amperes.

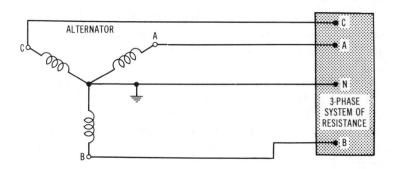

Fig. 6-8. Wye-connected generator and load.

Q6-9. What is the line voltage in the system in Fig. 6-8?

Q6-10. What is the phase voltage?

Q6-11. What is the line current?

Q6-12. What current flows through the alternator windings?

Q6-13. What is the power used in each single-phase circuit in this system?

Q6-14. What is the total power in all three phases?

Your Answers Should Be:

A6-9. The line voltage is **208 V.**

A6-10. The phase voltage is the line voltage divided by 1.73.

$$\frac{208 \text{ V}}{1.73} = 120 \text{ V}$$

A6-11. The line current is **10 amperes.**

A6-12. The line current of **10 amperes** flows through the alternator windings.

A6-13. The power in each single-phase circuit is E × I × power factor.

$$120 \times 10 \times 1 = 1200 \text{ watts}$$

A6-14. Since there are three single-phase circuits, the total three-phase power is 3 × 1200 = **3600 watts.**

Balanced Loads

In the preceding example (Fig. 6-8), all three line currents (and all phase currents) were equal. When **all 3 currents** in the lines **are equal** and are 120 electrical degrees apart, the load is said to be **balanced.**

When a load is balanced, the three currents meet at the neutral point of the load and at the neutral point of the generator and cancel. This cancelling effect is due to the phase relationships that the currents have to each other. As long as all three currents are equal, there will be no current present in the neutral wire.

If the load is not balanced, current will flow in the neutral wire. The amount depends on the amount of current in each line and the phase relationships of the line currents to each other. It is always good practice to balance the load in any three-phase system whenever it is possible to do so.

Power

You have found the power in a three-phase wye-connected system with a balanced resistive load. This was done by multiplying phase current times phase voltage for each phase and, then, adding to find the total power.

Of course, if the load is inductive or capacitive, you must also multiply by the power factor. The formula for finding the power in **any kind of balanced load** in a wye-connected three-phase system is:

$$P_T = I_p \times E_p \times 3 \times \cos \theta$$

where,
    P_T is the total power,
    I_p is the phase current,
    E_p is the phase voltage,
    $\cos \theta$ is the power factor.

There is also another way of finding power in a three-phase wye-connected system with a balanced load. Since the phase voltage is always equal to the line voltage divided by 1.73, then:

$$P_T = \frac{\text{line voltage} \times \text{line current} \times 3 \times \text{power factor}}{1.73}$$

Note that line current is the same as phase current and, also, that 1.73 is the square root of 3. Therefore, $\frac{3}{1.73} = 1.73$. When the formula is simplified, it becomes:

$$P_T = I_L \times E_L \times 1.73 \times \cos \theta$$

where,
    P_T is the total power,
    I_L is the line current,
    E_L is the line voltage,
    $\cos \theta$ is the power factor.

Q6-15. Is the load balanced in a three-phase wye-connected system whose line currents are 10 amperes, 10 amperes, and 15 amperes?

Q6-16. In the system in Question 6-15, will there be any current in the neutral wire?

Q6-17. A balanced wye-connected three-phase system has a line voltage of 208 V and a line current of 5 amperes. The power factor of the load is 0.8. What is the total power supplied to the load? What is the current in the neutral wire?

DELTA CONNECTION

The three-phase, **delta-connected** system gets its name from the appearance of the diagram of its connections. Delta is the name of the Greek letter Δ.

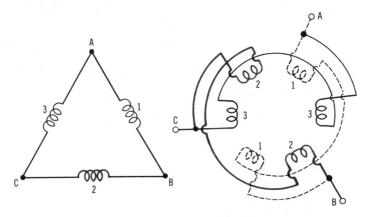

Fig. 6-9. Delta connection.

The windings in a delta system are connected end to end in a sort of loop. When the system is properly balanced, almost no current flows around the loop because of the phase differences of the voltages. The delta-connected system has no neutral point. Delta-connected systems are not usually grounded.

The wye connection is essentially a series connection. The phase voltages combine to produce a higher line voltage because the line voltage is developed across two windings in series. The phase current and line current are the same. The delta connection, however, is essentially a parallel connection. The line voltage is simply the voltage developed across an individual winding, so the line voltage is equal to the phase voltage. The currents from two windings combine to give the line current. The amount of the line current depends on both the amount of the phase currents and the phase difference between them. If the load is balanced and the voltages are equal, the line current is 1.73 times the phase current.

In a balanced delta-connected system, the power can be found by multiplying phase voltage times phase current times the number of phases times the power factor:

$$P_T = I_p \times E_p \times 3 \times \cos \theta$$

The line voltage is equal to the phase voltage. The phase current is equal to the line current divided by 1.73. This leads to the following formula:

$$P_T = I_L \times E_L \times 1.73 \times \cos \theta$$

where,

P_T is the total power,
I_L is the line current,
E_L is the line voltage,
$\cos \theta$ is the power factor.

Note that the two power formulas for delta-connected systems are the same as the two for wye-connected systems.

Q6-18. For the circuit shown in Fig. 6-10, what is the line voltage, line current, phase voltage, phase current, and total power?

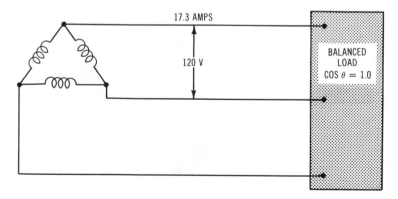

Fig. 6-10. Sketch for Question Q6-18.

POWER MEASUREMENT

So far, you have learned how power is calculated but not
how it is measured. In practical cases, all the information
needed to calculate power may not always be available. The
power factor of the load may be unknown, the load may not
be balanced, or the voltage may fluctuate. In many cases, it
may be necessary to measure power with a wattmeter.

Wattmeters have, as inputs, the factors one needs to know
in order to calculate power—voltage and current. In effect, a
wattmeter measures I and E and, then, calculates P mechanically. A wattmeter has a series connection to the line to measure current and a shunt connection to measure voltage. On
power circuits, the wattmeter is connected to the main lines
by a set of special instrument transformers that step down
the voltage and current to safe values.

The Three-Wattmeter Method

The three-wattmeter method is used primarily with three
phase four-wire circuits. It can measure power in both bal-

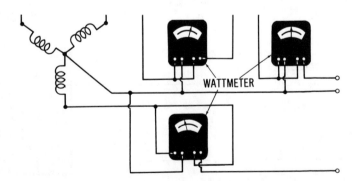

Fig. 6-11. The three-wattmeter method of measurement.

anced and unbalanced systems. In this simple method, three wattmeters are used to measure the power in all three phases at the same time. A typical connection circuit is shown in Fig. 6-11.

In the three-wattmeter method, each wattmeter measures a separate phase. The meter receives as inputs the phase volt-

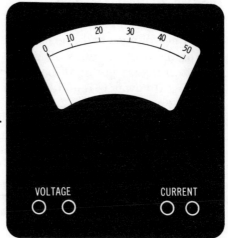

Fig. 6-12. A simple wattmeter.

age and phase current and gives a reading indicating phase power. In order to find the total power, add all three readings directly.

Q6-19. The power in a three-phase four-wire wye-connected system is measured using the three-wattmeter method. The individual meter readings for different loads are as shown in the following chart. Find the three-phase power values. Which load is balanced?

	W1	W2	W3
A	30W	40W	20W
B	20W	20W	20W
C	60W	60W	20W

The Two-Wattmeter Method

The two-wattmeter method of measurement (Fig. 6-13) employs line voltages and line currents and is suitable for either wye or delta connections. The two-wattmeter method can be used to measure power in both balanced and unbalanced circuits. Total power is found by adding the two readings shown

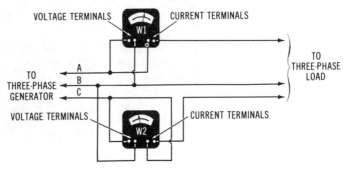

Fig. 6-13. Two-wattmeter method of measuring power.

on the wattmeters. If the power factor of the load is less than 0.5, the wattmeter readings must be subtracted to obtain the total power. The sum (or difference) of the meter readings gives the total power because of the phase relationships of the voltages and currents in the three-phase system.

You can make a test to determine whether the power factor is less than 0.5. Connect the wattmeters as shown in Fig. 6-13 and so that both give an up-scale reading. (If the pointer of one of the wattmeters moves in the wrong direction, the voltage leads of this wattmeter must be reversed to obtain a reading.) Then, temporarily move the voltage lead of W1 from line B to line C. (Or move the lead of W2 from line B to line A.) If the wattmeter reading moves down-scale, the power factor is less than 0.5, and the smaller reading must be subtracted from the larger reading.

The One-Wattmeter Method

If the three-phase load is **balanced,** one wattmeter can be used to measure the total power as shown in Fig. 6-14. A read-

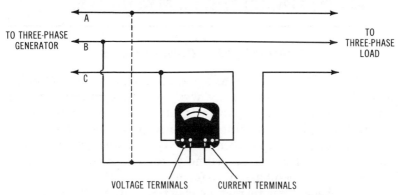

Fig. 6-14. One-wattmeter method of measuring power.

ing is taken with one voltage lead connected to line B. The lead is then moved to line A and another reading is taken. The sum of the readings gives the total power. A power factor less than 0.5 produces the same result in the one-wattmeter system as in the two-wattmeter system. Of course, switches can be used to change the voltage leads in either the one- or two-wattmeter method. Wattmeters that have negative-zero-positive scales may also be used in both methods.

Q6-20. The two readings recorded when measuring the power by the two-wattmeter method are +20 W and +40 W. Find the total three-phase power.

Q6-21. The two readings recorded when measuring the power in a circuit using the two-wattmeter method and +40 W and −15 W, respectively. Find the total three-phase power.

Q6-22. Four sets of readings recorded using the one-watt-meter method are:

A. +40, −20
B. +40, +40
C. +15, +20
D. −15, +30

Find the total power for each set of values.

TRANSFORMER CONNECTIONS

Wye and delta connections are used in many types of equipment. Three-phase motors are wye- or delta-connected, as are alternators. Another important type of a three-phase device is the three-phase transformer. Since power is often transmitted three-phase and since its voltage is transformed up and down at several points according to the needs of the distribution system, three-phase transformers are quite important.

Three-phase transformers consist of either three single-phase transformers connected together or a single-core three-phase winding, as shown in Fig. 6-15.

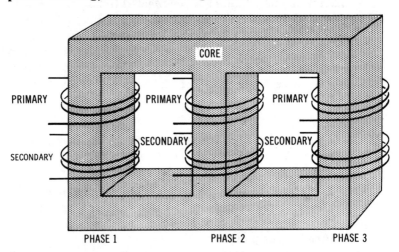

Fig. 6-15. A three-phase single-core transformer.

The primaries and the secondaries can be connected in any combination of delta and wye. For example, the primaries can be wye-connected and the secondaries can be delta-connected, or vice versa. Or, both the primaries and the secondaries can be wye- or delta-connected.

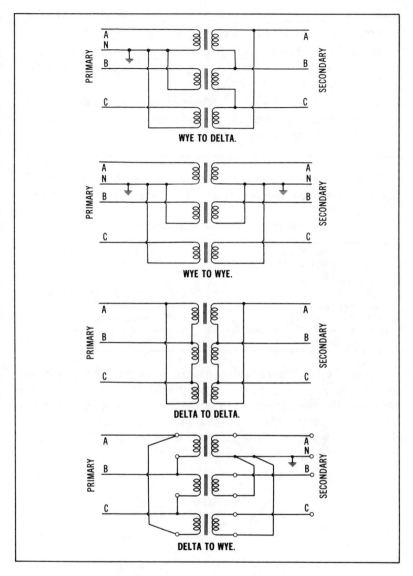

Fig. 6-16. Three-phase transformer connections.

Q6-23. Name the four ways in which transformers can be connected in a three-phase system.

Q6-24. What are the two kinds of three-phase transformers?

Both three-phase transformers and three single-phase transformers have their advantages. These are listed in Table 6-1.

Table 6-1. Transformer Advantages

Three-Phase Transformer	Three Single-Phase Transformers
Lower initial cost. Higher efficiency. Less total weight. Less total floor space. Lower installation and transportation cost.	Cheaper spare parts. A single unit can be replaced in case of trouble. Lower repair cost. More voltage flexibility.

Overall, the three-phase transformer is considered a better
unit in most situations. One minor advantage of using three
single-phase transformers in a delta connection is that it is
possible to remove one of the transformers altogether and still
have the system operate at reduced capacity. This arrangement is called an **open delta** configuration.

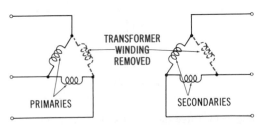

Fig. 6-17. Open-delta configuration.

When using wye-to-delta or delta-to-wye connections, there
is one important fact to remember. Line voltage and phase
voltage are equal in the delta connection and line voltage is

1.73 times the phase voltage in the wye connection. These facts can be used to get an additional step up or step down of voltage without adding extra turns to the transformers.

For example, suppose you are using a group of single-phase transformers with turns ratios of 1 to 5 to step up a three-phase voltage of 100 volts (line voltage). If the primaries of the transformers are delta-connected, the primary phase voltage is 100 volts. The phase voltage at the secondary will be 500 volts, but if the secondary is **wye-connected,** the line voltage will be 1.73 × 500 = 865 volts. If both sets of windings had been connected in the same way (both wye or both delta), the secondary line voltage would have been only 500 volts.

Q6-25. What is the secondary line voltage in the circuit of Fig. 6-18?

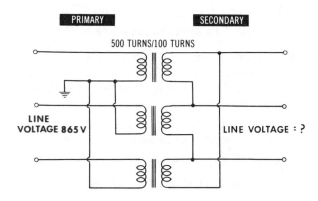

Fig. 6-18. Sketch for Question Q6-25.

Q6-26. Draw a diagram showing three single-phase transformers connected in a wye-to-delta arrangement.

Q6-27. A delta-to-wye three-phase transformer has a turns ratio of 10 to 1 and is used as a step-down transformer. If the input line voltage is 1000 V, what is the output line voltage?

Q6-28. What is one disadvantage of using the open-delta connection?

Your Answers Should Be:

A6-25. The phase voltage of the wye-connected primary is $865/1.73 = 500$ V. The phase voltage in the secondary is 100 V. Since the secondary is delta-connected, the **line voltage is also 100 V.**

A6-26.

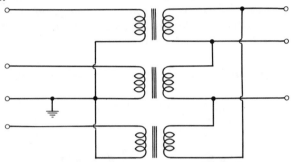

Fig. 6-19. Sketch for Answer A6-26.

A6-27. The primary phase voltage is 1000 V. The secondary phase voltage is $1000/10 = 100$ V. Since the secondary is wye-connected, the **secondary line voltage is 173 V.**

A6-28. In the open-delta connection, the transformers must be operated at **reduced capacity.**

WHAT YOU HAVE LEARNED

1. The power for most household and industrial use is generated and distributed as three-phase power.
2. Three-phase power is usually generated in hydraulic or steam power plants.
3. In hydroelectric plants, falling water drives a generator at low speed.
4. In steam plants, steam drives a turbine at high speed. In steam plants, the turbine, alternator, and dc exciter are arranged horizontally.
5. Wye and delta connections make it possible to transmit three-phase power with four, or even three, conductors.
6. In the wye connection, one end of each winding is connected to a common point which is often grounded.

7. In the wye connection, the voltage between the lines is called line voltage and is 1.73 times the voltage generated in any phase winding.

8. The voltage in a phase winding is called phase voltage, and appears between a line and the neutral wire in the wye connection.

9. In the delta connection, the windings are connected to form a closed loop. No current flows around this loop.

10. In the delta connection, line voltage and phase voltage are equal.

11. In the delta connection, line current is 1.73 times the phase current.

12. In any three-phase system, the total power is equal to the sum of the powers in all three phases.

13. The power in a three-phase system can also be found from line voltages and currents by using the formula:

$$P_T = 1.73 \, I_L \, E_L \cos \theta$$

14. Power in three-phase systems can be measured using one, two, or three wattmeters.

15. In the three-wattmeter method, power is simply measured in each phase. This method is used primarily with three-phase four-wire wye-connected systems.

16. In the one-wattmeter and two-wattmeter methods, two readings must be either added or subtracted (depending on the power factor) to find the power in the system.

17. Transformers can be connected in wye or delta patterns.

18. The primary and secondary of a three-phase transformer need not be connected in the same pattern.

19. By connecting the primary of a transformer in a wye configuration and the secondary in delta, or vice versa, it is possible to have a greater change in voltage than is produced by the turns ratio of the transformer alone.

20. Three-phase transformers can be three-phase units or combinations of three single-phase transformers.

21. Three single-phase transformers can be connected in an open-delta configuration.

7

Power Converters

what you will learn

There are a number of devices that convert one type of power to another. You will now find out how direct current can be converted to alternating current, and vice versa. You will learn how ac frequency can be changed and dc voltages stepped up or down. You will also learn how to draw schematics for some of these changes and how to choose the correct device for a particular application.

THE NEED FOR CONVERTERS

Electrical energy is generated and transmitted primarily as three-phase ac current, usually at 60 Hz and at very high voltages. The voltages are transformed to lower values before being used. Sometimes only one phase is used for a particular piece of equipment. The most common type of power supplied to homes is single-phase alternating current.

Occasionally, direct current is required at a remote location in rather large quantities. **Converters** provide a means of changing available ac currents to dc currents. Sometimes the supply source is dc current when ac currents are needed. A different kind of converter meets this need. Sometimes special frequencies of alternating current are required or high dc voltages are needed. Converters can provide these also. The word **converter** actually includes any of the following changes in electrical form: ac to dc, ac to ac of a different frequency, dc to ac, dc to higher-voltage dc, dc to lower-voltage dc, etc.

DC-TO-AC CONVERTERS

The conversion of dc to ac usually takes place when a dc source is available and a rather specialized ac frequency is required. The two basic methods of converting dc to ac are by means of a motor-generator (M-G) set and by vibrator action.

Motor Generators

In the simplest form of M-G set, a dc **motor** drives an ac **generator.** Depending on the type of ac generator chosen, a dc-to-ac converter may deliver single-phase, two-phase, or three-phase ac power. Most generators in this type of converter are three-phase synchronous units.

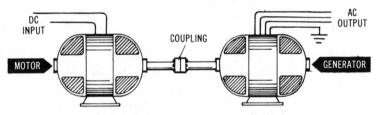

Fig. 7-1. A dc-to-ac motor-generator set.

A dc motor that is chosen as the prime mover in an M-G set must have speed characteristics that are very constant. If the speed of the motor varies considerably as the load increases, the frequency of the ac generator will vary accordingly. (Frequency is directly proportional to the speed of the prime mover.) One suitable type of dc motor is the shunt motor, although compound motors are also used to a great extent. A series motor could not normally be used because of its poor speed control.

Vibrators

Vibrators are used in dc-to-dc converters and in dc-to-ac converters but usually are able to deliver only very small amounts of power. Fig. 7-2 shows a simple vibrator circuit. (The filtering components have been deliberately omitted so that the reader may better understand how the circuit works.)

When switch S is closed, current will flow through the magnet coil. This action causes the coil to attract the reed and thus close contact A. When the reed touches contact A, it short-circuits the magnet coil, thus releasing the reed from its closed position. The reed is released and it springs back. Its inertia sends it into contact with point B. By this time, the magnet coil again has current flowing through it, so it pulls the reed back to point A, and the cycle repeats itself.

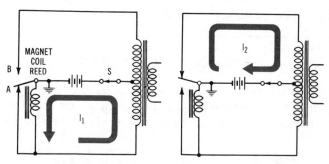

(A) Current through lower winding.   (B) Current through upper winding.

Fig. 7-2. A vibrator circuit.

The vibrating reed causes the current to flow first through the lower section of the transformer primary winding and, then, through the upper section. This is very much like having an alternating current flowing in the primary. This action causes an ac voltage to be induced in the secondary of the transformer. The voltage is not a sine wave, however, but has a generally rectangular shape. Filtering action can smooth out the waveform.

The magnitude of the induced voltage depends on the turns ratio of the transformer. The frequency of the voltage depends on the speed with which the reed changes from point A to point B. The speed, in turns, varies with the weight of the reed and the strength of the magnetic field.

Q7-1. Do you think a special converter would be used to convert low-voltage ac to high-voltage ac? How would this be done?

Q7-2. Why is it necessary to use a constant-speed motor in a dc-to-ac motor-generator set?

Q7-3. The moving part of a vibrator is a _ _ _ _ _ _ _ _ _ _ _ _ _ .

DC-TO-DC CONVERTERS

A simple means of obtaining a higher dc voltage from an existing dc source is to use an M-G set in which both the motor and the generator are dc machines. A shunt or compound dc

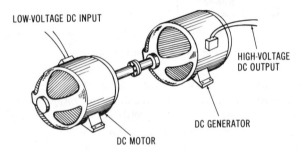

LOW-VOLTAGE DC INPUT

HIGH-VOLTAGE
DC OUTPUT

DC GENERATOR

DC MOTOR

Fig. 7-3. A dc-to-dc motor-generator set.

motor is best because it has a relatively constant speed for varying loads. The output voltage of the generator depends on the speed. A shunt or compound dc generator is best suited for use in M-G converters because it has a good voltage regulation for varying loads.

When only a small dc output is required, a vibrator can be successfully employed. The vibrator converts the direct current to higher-voltage alternating current. Then, the ac current is converted back to dc current at this higher voltage. A typical circuit is shown in Fig. 7-4.

The advantage of the dc-to-dc vibrator converter is that the ac output (produced by the parts enclosed in the dashed line in the illustration of Fig. 7-4) can be converted to a dc voltage that is higher than that of the battery. The means by which ac can be converted back to dc will be discussed in the follow-

ing section. The device shown in Fig. 7-4 is a **full-wave rectifier**, a type of electronic converter. The dc-to-ac devices discussed so far are often called **inverters**. This distinguishes
them from the more common ac-to-dc devices that are called
converters.

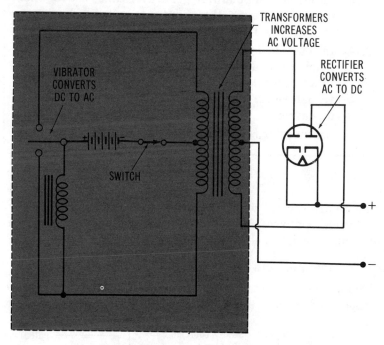

Fig. 7-4. A dc-to-dc vibrator converter.

Q7-4. What sort of dc motor would be best suited for use
in a motor-generator set that is used as a dc-to-dc
converter? Why?

Q7-5. What sort of dc generator would be best suited for
this application? Why?

Q7-6. In one device for raising the voltage of small
amounts of dc power, the dc current is converted
to _ _ _ _ _ _ _ _ _ and, then, the voltage is raised
by a _ _ _ _ _ _ _ _ _ _ _ _. The ac current is then
converted to _ _ _ _ _ _ _ _ _ .

Q7-7. An inverter is a device for converting _ _ to _ _ .

AC-TO-DC CONVERTERS

There are several ways of converting ac current to dc current, such as any of the following: motor-generator set, synchronous converter, electronic rectifier, or contact rectifier.

Motor-Generator Sets

The use of M-G sets as ac-to-dc converters is similar to the method used in inverters. An ac motor drives a dc generator.

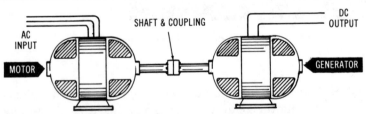

Fig. 7-5. An ac-to-dc motor-generator set.

It is customary to use three-phase motors of either the induction or synchronous type. The induction motor has the advantage of being lower priced and more rugged in construction than the synchronous motor. The induction motor can be wound for voltages as high as 13,500 volts, thus eliminating the need for step-down transformers in many cases. Synchronous motors have the advantage of constant speed and a higher power factor. In large converter installations, the synchronous motor is usually preferred. By overexciting a synchronous motor, the overall power factor of the system can be improved. The overexcited synchronous motor becomes a synchronous capacitor. (That is, it does the job of a huge capacitor.)

The Synchronous Converter

The synchronous converter is sometimes called a **rotary converter**. It changes alternating current to dc current. The synchronous converter consists of a dc field and a dc armature that is equipped with slip rings in addition to a commutator (Fig. 7-6); consider it as a synchronous motor whose armature is equipped with a commutator to provide a dc output.

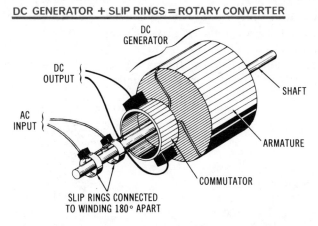

DC GENERATOR + SLIP RINGS = ROTARY CONVERTER

Fig. 7-6. Armature of a synchronous converter.

Under ordinary conditions, an alternating voltage is connected to the motor portion of the converter through slip rings, causing the machine to rotate at a synchronous speed like an ac synchronous motor. At the same time, the machine also acts like a dc generator. The dc output is taken from the commutator by brushes and can be used to provide current for the field winding. This dc output can, in addition to energizing the field winding, furnish dc current to an external load.

Q7-8. A synchronous converter has an armature equipped with _ _ _ _ _ _ _ _ _ and a _ _ _ _ _ _ _ _ _ _.

Q7-9. The dc output is taken from a synchronous converter through the _ _ _ _ _ _ _ _ _ _ brushes.

Electronic Rectifiers

The simplest and most common type of electronic rectifier is diode vacuum tubes (electron tubes). The principles of electron-tube operation are covered more thoroughly in Volume 3 of this series.

A vacuum-tube diode usually consists of three metal parts enclosed in a glass bulb from which the air has been removed to create a vacuum. The parts are the plate, cathode, and heater. In some tubes, the filament also serves as the cathode; this is called a directly heated cathode.

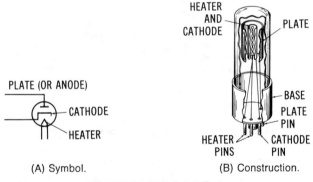

(A) Symbol.   (B) Construction.

Fig. 7-7. Vacuum-tube diode.

In a vacuum-tube diode, heating the cathode drives electrons off the cathode. If a positive voltage (with respect to the cathode) is applied to the plate, the plate will attract the electrons emitted from the cathode. The movement of electrons is, in effect, a current flow. If the plate is more negative than the cathode, no current flows since there is no positive charge to attract the electrons. The vacuum-tube diode thus acts as a valve. It conducts in one direction but not in the other.

Look at the circuit diagram in Fig. 7-8. During the positive half cycle of the ac input, the plate is positive with respect to the cathode, and current flows. (A voltage drop appears across the load resistor, but it is a little less than the input voltage.) During the negative half cycle, the plate is negative with respect to the cathode and no current flows.

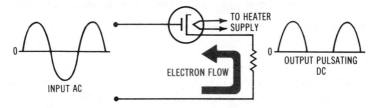

Fig. 7-8. Half-wave rectifier.

The device just described is called a **half-wave rectifier** because only half of the ac sine wave appears in the output. Although this current never reverses direction, it is a very rough pulsating dc current which is not satisfactory for most purposes. A smoother dc current can be obtained from a **full-wave rectifier** (Fig. 7-9).

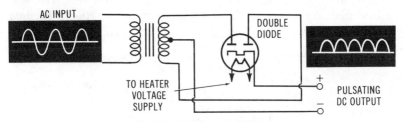

Fig. 7-9. Full-wave rectifier.

Both halves of the ac sine wave appear in the output of the full-wave rectifier. The output is still a pulsating direct current, but it is smoother than the half-wave output. With proper filtering, the output of a full-wave rectifier can be smoothed further into a steady dc current.

Q7-10. A vacuum-tube diode conducts only when the plate is _ _ _ _ _ _ _ _ with respect to the cathode.

Q7-11. The output from a _ _ _ _ - _ _ _ _ rectifier is easier to smooth than the output from a _ _ _ _ - _ _ _ _ rectifier.

Contact Rectifiers

Contact, or **barrier-layer,** rectifiers are devices which permit current flow in one direction only. (Actually, they present a very high resistance in the reverse direction.) They produce about the same result as a vacuum-tube diode. Two types of metallic barrier-layer rectifiers that were once very common are copper-oxide and selenium rectifiers.

The **copper-oxide** rectifier (Fig. 7-10A) is produced by heating a copper disc to a high temperature and, then, quenching it in water. This produces a thin layer of red cuprous oxide sandwiched between the copper disc and a thick outer layer of green cupric oxide. The cupric oxide is then removed and a lead disc is pressed against the cuprous oxide. Electrons flow more easily from lead to copper than from copper to lead.

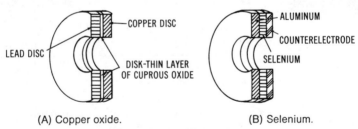

(A) Copper oxide. (B) Selenium.

Fig. 7-10. Contact rectifiers.

The **selenium** rectifier (Fig. 7-10B) is made by depositing a layer of selenium on an aluminum plate. An alloy with a low melting point is then sprayed onto the selenium surface. This alloy is called the **counterelectrode.** The current-blocking layer is the surface between the selenium and the alloy. Electrons flow through this surface easily in one direction but not in the other.

The type of rectifier most widely used now is the **semicon-**

ductor rectifier. The rectifying action is produced by the junction of p-type and n-type materials. Electrons flow easily from the n-type material to the p-type material, but not in the reverse direction. You can learn more about the theory of semiconductor materials by studying Volume 3 of this series.

Small rectifier cells are often called **rectifier diodes** since they perform the same rectifying action as the diode vacuum tube. If large quantities of electricity must be rectified, **rectifier stacks** are used. These stacks are composed of a number of rectifier elements that are assembled and connected together.

The symbol for a rectifier is shown in Fig. 7-11. The arrowhead points **against** the direction in which electrons move through the rectifier.

Fig. 7-11. Contact rectifier symbol.

The rectifiers shown so far have had single-phase inputs. Three-phase current can also be rectified, as shown by the illustration in Fig. 7-12.

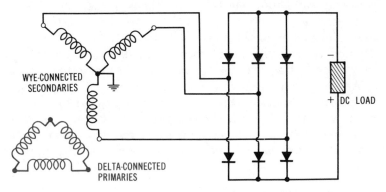

Fig. 7-12. Three-phase rectifier circuit.

Q7-12. Name three types of rectifier cells.

Q7-13. Draw a schematic of a full-wave rectifier using two rectifier cells.

Q7-14. What property of rectifiers makes them useful in converting alternating current to direct current?

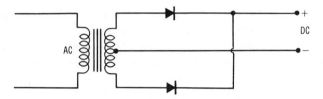
SOLID-STATE CONVERTERS

The conversion of a dc voltage into another dc voltage, or a dc voltage to an ac voltage, is now being done mainly with the use of solid-state electronic methods rather than using motor generators, vibrators, and vacuum-tube rectifiers. Of course, the earlier methods are still widely in use in existing equipment so a knowledge of those methods is still important.

Now, let's look at a couple of examples of solid-state converters. An electronic dc-to-ac converter can consist simply of a transistorized oscillator whose output is connected to a transformer with the required step-up ratio necessary to provide the needed ac voltage value. If dc-to-dc voltage conversion is the objective, all that you need to do is add a rectifier to the output.

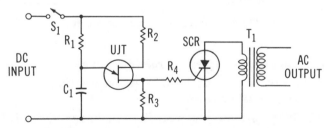

Fig. 7-13. Triggering an SCR with a UJT oscillator.

Since transistors are not able to handle the high currents needed for some applications, a UJT oscillator is often used to provide the triggering for the gate of an SCR (Fig. 7-13). The UJT oscillator turns the SCR on and off at a triggering rate that is determined by the time constant of an R1–C1 combination located in the UJT gate circuit. This produces a high-power ac voltage through the SCR. The ac voltage that is produced may then be applied to the windings of a transformer to step up or step down the voltage to any desired ac voltage.

The UJT circuit shown in Fig. 7-13 operates as follows. When switch S1 is closed, capacitor C1 starts charging through resistor R1 to the full dc voltage provided by the power source. At first, the voltage at the emitter of the UJT is zero as all the current goes to charging the capacitor. Then, as the capacitor becomes charged, the voltage across it increases to a point where the emitter voltage causes the UJT to suddenly fire. This shorts out the capacitor and the cycle starts to repeat. With each cycle, a voltage pulse which acts as a trigger for the SCR is developed across resistor R3.

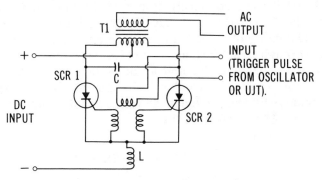

Fig. 7-14. A two-SCR converter circuit.

A circuit containing two SCRs is shown in Fig. 7-14. Pulses can be provided to the trigger input circuit by either a tran-

Q7-15. A disadvantage of using a transistor oscillator and rectifier as a dc-to-dc converter is that the output _ _ _ _ _ _ is limited.

Q7-16. Instead of using a transistor oscillator for dc-to-dc or dc-to-ac conversion, a _ _ _ oscillator is often used to trigger an _ _ _ .

sistor oscillator or a UJT circuit like the one just described. Direct current passes through each SCR, alternately, as each SCR is triggered. The cathodes of the SCRs are connected in parallel and, also, through inductor L to the minus (−) side of the dc input line. Taking SCR 1 first, we find that if its gate is triggered, electrons will flow from the minus (−) side of the dc input, through coil L and SCR 1, through the left side of transformer T1 to the midtap (from left to right), and back to the positive side of the dc input. Then, when SCR 2 is triggered and it conducts, electron flow is through SCR 2 to the right side of T1 and to the transformer midtap (from right to left). The firing of SCR 1 and SCR 2, and their reverse currents through transformer T1, causes a pulsating voltage across the primary of T1 and an ac voltage in its secondary. The ac output at the secondary winding can then be rectified to provide a different value of dc voltage as determined by the turns ratio of transformer T1.

FREQUENCY CONVERTERS

The last type of converter to be discussed is the frequency converter. The simplest way of changing frequency is to provide a motor-generator set in which the motor is an ac (preferably synchronous) unit and the generator is an ac synchronous unit, but with a different number of poles.

Look at the drawing in Fig. 7-15. If the machine is a generator, the emf induced in each winding goes through a complete cycle every time two unlike poles pass the winding. In the illustration, this happens once for every revolution of the rotor. If the machine is a synchronous motor, the rotor will make a complete revolution for each cycle of current in the windings.

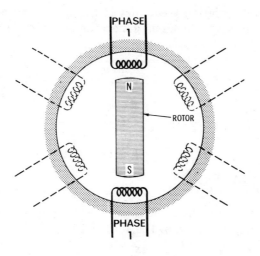

Fig. 7-15. A two-pole machine.

Now look at Fig. 7-16. This machine has four poles on the rotor. **Four poles** pass each winding during each revolution. Therefore, there are **two complete cycles** of current in each winding for each revolution of the rotor. A machine with **eight poles** will have **four complete cycles** of current in each winding for each revolution of the rotor, and so on.

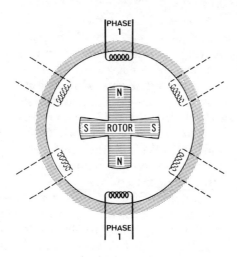

Fig. 7-16. A four-pole machine.

The formula relating frequency to the speed and to the number of poles in a synchronous generator or motor is:

$$S = \frac{120 \times f}{p}$$

or,

$$f = \frac{S \times p}{120}$$

where,
   S is the speed (rpm),
   f is the frequency (Hz),
   p is the number of poles.

Thus, if the motor in an M-G set is a two-pole machine operating at 60 Hz, the speed of the shaft will be:

$$S = \frac{120 \times f}{p}$$
$$= \frac{120 \times 60}{2}$$
$$= 3600 \text{ rpm}$$

If this motor is used to drive a generator having six poles, the output frequency will be:

$$f = \frac{S \times p}{120} = \frac{3600 \times 6}{120} = 180 \text{ Hz}$$

The ac-to-ac motor-generator set is probably the simplest frequency-converting device. However, a number of other devices can be used for this purpose.

The **induction generator** discussed in Chapter 4 is basically a frequency-converting device. As you remember, it can be used to decrease the frequency if the rotor is driven in the same direction as the rotating magnetic field. If the rotor is driven in the opposite direction, the frequency is increased.

Higher frequencies for small amounts of current can be obtained using **vibrators** operated by ac current instead of direct current. Very small quantities of power at almost any frequency can be developed with an electronic **oscillator**. Since this type of device is used mainly in electronic work, it is discussed in Volumes 3 and 4 of this series. The oscillator is basically a dc-to-ac converting device.

Q7-17. Find the output frequency of the M-G set shown in Fig. 7-17.

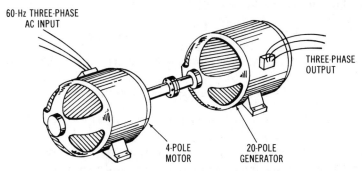

60-Hz THREE-PHASE
AC INPUT

THREE-PHASE
OUTPUT

4-POLE
MOTOR

20-POLE
GENERATOR

Fig. 7-17. Sketch for Question Q7-17.

Q7-18. Does a four-pole generator have to turn faster or slower than a two-pole machine to generate the same frequency?

Q7-19. Find the output frequencies when the motor input frequency is 60 Hz and the poles are as shown.

	Motor Poles	Generator Poles
A	2	8
B	4	10
C	16	4
D	10	70

Q7-20. What device having no moving parts would you use to convert a moderate amount of alternating current to direct current?

Q7-21. What type of device would you use to obtain heavy dc currents from alternating current?

Q7-22. What type of device would you use to obtain a small ac current from direct current?

Q7-23. How could you step up the voltage of a small amount of dc power?

Q7-24. The word inverter is often used to refer to a device that converts _ _ current to _ _ current.

Your Answers Should Be:

A7-17. The speed of the motor is:

$$S = \frac{120 \times f}{p} = \frac{120 \times 60}{4} = 1800 \text{ rpm}$$

The frequency of the generator is:

$$f = \frac{S \times p}{120} = \frac{1800 \times 20}{120} = 300 \text{ cps}$$

A7-18. A four-pole generator turns at **half the speed** of a two-pole machine to generate the same frequency.

A7-19. (A) 240 Hz, (B) 150 Hz, (C) 15 Hz, (D) 420 Hz.

A7-20. A **rectifier** has no moving parts and can be used to convert a moderate amount of alternating current to direct current.

A7-21. **An M-G set with a synchronous motor and a shunt or compound dc generator** will convert heavy ac currents to direct current.

A7-22. **A vibrator** will provide a small ac current from a dc input.

A7-23. **A vibrator combined with a transformer and a rectifier** can step up dc voltages.

A7-24. The word inverter is often used to refer to a device that converts **dc** current to **ac** current.

WHAT YOU HAVE LEARNED

1. Power converters are used to change one type of power to another. For example, they change ac to dc, dc to ac, ac to ac of a different frequency, or dc to a higher-voltage dc.

2. Direct current can be converted to alternating current with the use of a motor-generator set or by using a vibrator.

3. An M-G set consisting of a dc motor and ac generator can convert dc to ac current. The best dc motor for this purpose is a shunt or compound type that has nearly constant-speed characteristics.

4. A vibrator can convert small amounts of dc current to ac current through the action of a vibrating reed.

5. Vibrators require a filtering circuitry to convert the square-wave output into a smoother sine-wave alternating current.

6. Vibrators can also be used to convert low-voltage direct current to high voltages. This is done by converting the low-voltage dc input to a high-voltage ac output with a vibrator and transformer and, then, using a rectifier to change it back to dc current.

7. Alternating current can be converted to dc current with the use of M-G sets, synchronous converters, electronic rectifiers, and contact rectifiers.

8. M-G sets that change ac currents to dc currents usually use a synchronous or induction motor and a shunt or compound generator.

9. A synchronous converter is a machine whose armature is equipped with slip rings and a commutator. It converts alternating current to direct current.

10. One type of electronic rectifier is a diode vacuum tube. An electronic rectifier allows current to pass in one direction but not in the other.

11. A contact rectifier operates similar to a vacuum-tube rectifier. Contact rectifiers have a high resistance in one direction and a low resistance in the other.

12. The most common types of contact rectifiers are: copper-oxide, selenium, and semiconductor.

13. Electronic and contact rectifiers can be connected in either half-wave or full-wave arrangements. A half-wave rectifier converts only half of the ac sine wave to direct current; a full-wave rectifier converts both halves of the ac sine wave to direct current and gives a smoother dc output.

14. Ac frequencies can be converted by the use of an M-G set consisting of a synchronous motor and a synchronous generator with a different number of poles.

15. Vibrators and induction generators are also used to convert the frequency of ac currents.

16. Dc-to-dc and dc-to-ac conversion techniques now often use solid-state devices such as UJTs and SCRs.

8

Servo
Control Systems

what you will learn

In this chapter, you will learn how a servo control system can move and precisely position a heavy load through application of a small input signal. You will become acquainted with the basic principles of servo and synchromechanisms and learn how these devices can be used to control the direction and amount of rotation of an electric motor. Also, you will discover how synchromechanisms can be used to transmit data from one location to another.

WHAT IS A SERVO CONTROL SYSTEM?

There are many load-positioning tasks that man has neither the muscle, mental ability, nor desire to perform. For the accomplishment of such tasks, he uses a servo control system. A servo system can be used to move the heavy rudder of a ship, position the beam of large searchlights, or sight an observatory telescope which weighs several tons on a distant star.

Servo systems can be designed for automatic operation. These systems can be used with radar for the automatic tracking of targets, in weapons systems for the automatic aiming of guns or missile launchers, in aircraft for the automatic positioning of control surfaces, and in many other installations where automatic and precise control is desired.

THE SERVO PRINCIPLE

The terms **servo system, servomechanism,** and **servo** are often used interchangeably to identify a complete system or one of its parts. To prevent confusion, an individual device will be identified by its noun name, servomotor for example, and a working combination of these devices will be called a servo (or control) system.

The Basic Servo System

All servo systems perform two functions. These functions are called **input control** and **output control.** Fig. 8-1 shows a

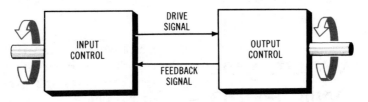

Fig. 8-1. Servo systems functions.

shaft feeding a rotational signal to the input control which causes the output control to rotate its shaft in the same direction. The turning of the output shaft moves a load to a desired position. The output control continuously feeds a signal back to the input to reveal the precise position (in a rotational direction) of the output shaft. If there is a difference between input and output shaft positions, the latter will continue to turn until the desired amount of rotation is attained. All servo systems operate in accordance with this principle.

Electrical, mechanical, hydraulic, or air-pressure devices may be used for control mechanisms or transmission of signals between the two control functions. Because of their economy, reliability of operation, and ease of control, most servo systems employ electrical or electromechanical devices. Transmission of signals between control devices is nearly always done electrically. This is particularly true when the input control is located at some point remote from the output control.

Since it reveals position difference between input and output, the feedback signal is frequently called an **error signal.** Sometimes the drive signal is called an **actuating signal.**

Basic Parts of a Servo System

A servo system may have as few as two or three devices or as many as a hundred or more. The number and variety of devices depend on the complexity of the system and the type of work it must perform. A modern radar, for example, must aim its antenna in elevation (up and down) and in bearing (left and right) as a result of manual or semiautomatic control. It must then automatically follow the target after it has been "locked-on." However, any servo system basically consists only of four functional devices as shown in Fig. 8-2.

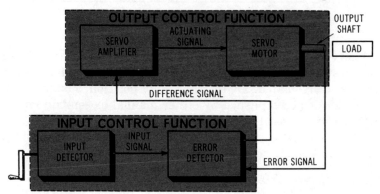

Fig. 8-2. Basic parts of a servo system.

The input detector senses any movement of the input shaft and sends a signal to the **error detector** where it is matched with an error signal that reveals the position of the output shaft. If the shafts are not in the same position, a difference signal is generated and, then, amplified by the **servo amplifier** to a value which will turn the **servomotor.** The motor will continue to turn until the difference between the input and error signals is zero. When this occurs, the two shafts will be in the same angular position.

Q8-1. What are the two signals which reveal the positions of the input and output shafts?

Q8-2. When a difference signal is generated, the input and output shafts (are, are not) in alignment.

Q8-3. What are the basic parts of the input and output control functions, respectively?

OPEN AND CLOSED SERVO SYSTEMS

A servo system may be defined as either open or closed. In an **open system,** the behavior or position of the output shaft is not automatically matched with the input position. Matching or correction is accomplished by either manual or semiautomatic methods. In a **closed system,** input and error signals are automatically compared to bring the input and output conditions into automatic alignment.

An Open Servo System

Fig. 8-2 shows an example of an open servo system. It is a steering system that could be used on a ship. The rudder turns in a direction and in an amount determined by the rotation of a wheel in the pilot house. Since the rudder is large and heavy, it is pivoted by a servomotor.

In this case, the man at the wheel is the input and error detector. He moves the wheel to maintain the ship on a specified course. He does this by watching the direction on a compass. When wind or ocean currents move the ship off course, he turns

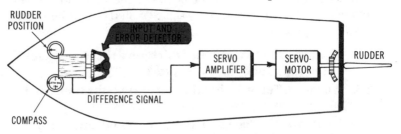

Fig. 8-3. An open servo control system.

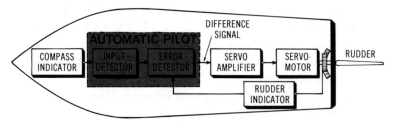

Fig. 8-4. A closed servo control system.

the wheel (and, consequently, the rudder) to bring the ship back on course. Because it is not automatic, the system is open.

A Closed Servo System

By replacing the man with an electric device, sometimes called an automatic pilot, the steering system can be made into a closed servo system as shown in Fig. 8-4.

When the ship drifts off course, the change in compass heading generates a voltage signal that is transmitted to the error detector. Since the rudder and heading (compass) positions no longer agree, a difference signal is sent to the servo amplifier and motor to turn the rudder an amount and in the direction required to bring the ship back on course. When the ship is back on course, the rudder will have been returned to its amidships (straight ahead) position.

If the ship is heading west (270° compass reading), for example, and the wind pushes it to a heading of 265°, there will be a 5° angular difference between the established positions of rudder and compass. A difference signal of an amount required to compensate for the deviation will be generated. The rudder will be turned to the right five degrees. Now the compass and rudder positions are the same, and the difference signal is zero. As the ship turns back to 270°, the rudder angle is decreased correspondingly and is returned to amidships when the ship heading becomes due west again. This system is a closed servo system.

Q8-4. A system which continuously self-compensates for differences between the input and output is a(n) _ _ _ _ _ _ servo system.

Q8-5. The difference signal will be zero when the _ _ _ _ _ signal is equal to the _ _ _ _ _ signal.

OPERATION PRINCIPLES

Fig. 8-5 is a simplified diagram of a steering servo system.

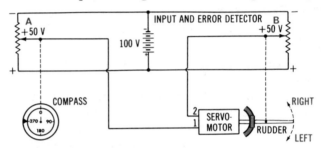

Fig. 8-5. A steering servo system.

The input and error detectors contain a **balanced potentiometer**—the resistances of both potentiometer legs are identical. Potentiometer arm A responds to the turning of a compass. Arm B follows the movement of the rudder. Arms A and B are connected to the servomotor at terminals 1 and 2, respectively. A 100-volt battery is connected across the potentiometer.

When the ship is steering course 270°, arm A is at the midpoint of the resistance. When the rudder is amidships, arm B is likewise at the midpoint of its resistance. Thus, both arms are selecting 50 volts. Since the voltage difference is zero, the motor will not turn.

If the ship swings 10° to the right, the compass will rotate to show a 280° heading. Arm A will be moved downward. Assume that it is now selecting +60 V. The rudder is still amidships and arm B is still selecting +50 V. Terminal 1 of the servomotor is now +10 V with respect to terminal 2. The motor is so wired that it will turn the rudder 10° left. When the rudder has reached the 10° position, arm B will have been moved to the +60 V position. The potential difference at the motor terminals is zero and the motor is stopped. This situation is shown in Fig. 8-6.

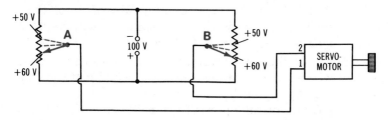

Fig. 8-6. Compass and rudder positions returning to zero.

The angle of the rudder now turns the ship to the left. Since 0° on the compass always points north, the compass scale remains steady, and the ship, in effect, turns under it. The heading of the ship begins to move from 280° back to the desired 270°. As it does, arm A makes a corresponding movement back to its midpoint. If, for example, arm A is selecting +58 V while arm B is still at +60 V, terminal 2 is now two volts positive with respect to terminal 1. The servomotor turns in a direction to decrease the rudder angle. Arm B follows the rudder movement, decreasing its distance from the center resistance point. As the ship continues to turn back to 270°, arm A is continuously leading arm B. Theoretically, arms A and B arrive at their midpoints simultaneously, and the difference signal becomes zero. This type of input control, however, is subject to overcompensation and will cause the ship heading to swing past the desired setting. Since the rudder will always follow, the ship will steer a winding course, weaving back and forth across the desired heading.

Q8-6. The resistances of a(n) _ _ _ _ _ _ _ _
_ _ _ _ _ _ _ _ _ _ _ _ _ are equal.

Q8-7. If arms A and B (in the example given on these two pages) are selecting ⅖ and ⅗ of their resistances, respectively (measured from the negative ends), what is the voltage across the motor terminals?

Q8-8. Assuming the rudder is amidships at the beginning of the conditions given in Question Q8-7, which way will it be turned?

SERVOMOTORS

The preceding example used a dc servomotor to position a load. If the voltage source had been alternating current, an ac servomotor could have been substituted. Requirements of a motor to be used in a servo system include the ability to drive a load in either direction and at varying speeds.

Dc Servomotors

Dc motors are selected for use in servo systems because of their high-stall torque characteristics and their ability to be operated at varying speeds. A **shunt-field motor,** the most predominant type of dc servomotor, is shown in Fig. 8-7.

Voltage across the shunt field is obtained from a source separate from the servo system. This produces a uniform magnetic field. Voltage fed to the armature is controlled by the output of the servo amplifier (dc). Speed of the motor is determined by the magnitude of the voltage difference between terminals 1 and 2. Direction of armature rotation is determined by the polarity of these voltages.

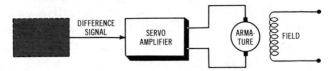

Fig. 8-7. A shunt-field dc servomotor.

Ac Servomotors

Where low power and low speed ranges are permissible in a servo system, an ac servomotor is used. A distinct advantage is the ability to use a comparatively simple servo amplifier when an ac power source is available.

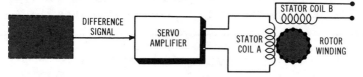

Fig. 8-8. A two-phase induction ac servomotor.

The diagram in Fig. 8-8 illustrates a **two-phase induction motor,** the type most frequently used in ac servomotor applications. Either a squirrel-cage or short-circuited rotor may be used. The two stator coils are physically displaced 90°. Coil B, connected to a separate excitation source, develops a steady magnetic field in the top and bottom poles. Coil A magnetizes the left and right poles and its magnetic field depends on the magnitude and phasing of the ac voltage obtained from the servo amplifier. As you learned in a previous chapter, the rotor will turn when the voltages applied to coils A and B are 90° out of phase.

The direction of rotor rotation is reversed when the servo amplifier reverses the direction of current through coil A. Speed of rotation is controlled by varying the strength of the magnetic field which, in this case, depends on the amount of current flowing through coil A.

In most installations, the value and direction of current flow is controlled by the characteristics of the difference signal fed to the servo amplifier. Since the difference signal will be an alternating current, the servo amplifier can be a relatively simple electronic amplifier. In small servo systems, the difference signal may be fed directly to the servomotor.

Q8-9. (Dc,Ac) servomotors are used in servo systems that require a high torque to move a load.

Q8-10. (Dc,Ac) motors have the wider range of speeds.

Q8-11. Dc servomotors are normally the _ _ _ _ _ -_ _ _ _ _ type and ac servomotors are the _ _ _ _ _ _ _ _ _ type.

SERVO AMPLIFIERS

A **servo amplifier**, as you recall, converts the difference signal into a voltage or current to drive a servomotor at a desired speed and in the correct direction. Dc servo amplifiers can be either electronic or electromechanical. Ac servo amplifiers are usually electronic.

Transistorized Dc Servo Amplifier

There are several varieties of electronic circuits used as dc servo amplifiers. Since nearly all difference signals are ac voltages, these circuits are designed to convert an ac signal into a dc output and to be sensitive to changes in phase. The circuit shown in Fig. 8-9 is typical.

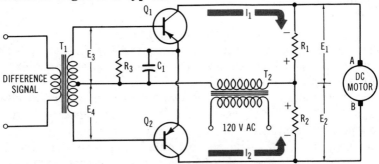

Fig. 8-9. A single-stage dc servo amplifier.

The servomotor used in such a circuit will probably be one with permanent-magnet poles. The secondary of T_2 applies an ac voltage to the collectors of transistors Q_1 and Q_2. As can be seen, the collector voltages are in phase. The difference signal enters through transformer T_1 and is applied to the two bases 180° out of phase. Capacitor C_1 and resistor R_3 form the common-emitter bias for both transistors. With no signal applied across the primary of T_1, each base is at the same potential and the collector currents are therefore equal. Voltage drops across the load resistors are equal and there is no difference of potential between points A and B of the motor armature. Therefore, the servomotor does not turn when there is no difference signal.

When a difference signal is applied across T_1, one of the bases will become positive with respect to the secondary center tap and the other base negative with respect to the same point. Since the collectors of transistors Q_1 and Q_2 receive power from the same line, one of the bases will be in phase with its collector voltage and the other out of phase. Assume that voltage E_3 is causing the base of Q_1 to become more positive (the collectors are going positive at the same time) while voltage E_4 is causing the base of Q_2 to go negative by the same amount. Collector current I_1 through Q_1 will increase and I_2 through Q_2 will decrease. Voltage E_1 will increase (negative to positive from top to bottom) and E_2 will decrease (negative to positive from bottom to top). The voltage appearing across the dc servomotor armature will be the algebraic sum of voltages E_1 and E_2. Since E_1 is greater than E_2, the armature voltage will be negative at point A with respect to point B. Current will flow through the armature from A to B, causing it to turn in a given direction.

When the difference signal reverses in phase, I_2 becomes greater than I_1. The difference between E_2 and E_1 causes an armature voltage which is negative at point B with respect to A. The armature will now turn in the opposite direction. Although the voltage across points A and B is pulsating dc, its waveform is sufficiently smoothed out to cause armature rotation. The speed of armature rotation is determined by the magnitude of the difference signal.

Q8-12. The collectors of Q_1 and Q_2 are always (in phase, 180° out of phase).

Q8-13. The bases of the two transistors are always (in phase, 180° out of phase).

Q8-14. The base of which transistor is always in phase with its collector?

Q8-15. _ _ _ _ _ and _ _ _ _ _ _ _ _ _ of the difference signal determine armature direction and speed, respectively.

Thyratron Dc Servo Amplifier

One way to obtain a greater amount of power from a dc servo amplifier is to use **thyratron** tubes. A thyratron is a gas-filled tube capable of passing eight or more amperes as compared to the milliamperes of a vacuum tube or small transistor.

A thyratron has a **firing potential** determined by its plate and grid voltages. Assuming a given bias voltage on the grid, plate voltage can be gradually increased until it reaches a value which overcomes the repelling effect of the grid. At this point, plate current immediately rises from zero to a value determined solely by the plate voltage. The grid no longer has control over the current. The usual method of interrupting the plate current is to shut off the plate voltage. A typical thyratron servo amplifier is shown in Fig. 8-10.

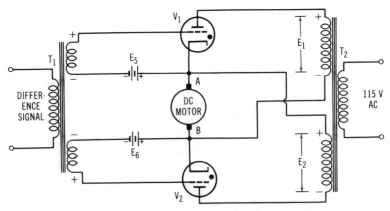

Fig. 8-10. A bidirectional thyratron dc servo amplifier.

Transformer T_2 is connected so that the plate voltages of V_1 and V_2 are 180° out of phase. Voltages E_5 and E_6 place a bias on each grid of sufficient value so that no plate current will flow when the difference signal is zero. Transformer T_1 is connected so the grid voltages are in phase with each other as shown in the diagram. Assume a difference signal on T_1 that causes the grids to swing positive while the plate voltage of V_1 is going positive. Tube V_1 will conduct. Current will flow in the secondary of T_2, through the dc servomotor from point B to A, and back to the cathode of V_1. During this half cycle, the plate of V_2 is negative and the tube will not conduct. When the plate of V_2 becomes positive during the next half cycle, its grid signal is going negative and prevents the thyratron from conducting.

When the phase of the difference signal is reversed, the grid of V_2 is going positive at the same time as its plate. Current flows through the servomotor in the opposite direction, from point A to B, causing it to rotate in the other direction. As in the vacuum-tube amplifier, the direction of motor rotation is determined by the phase of the difference signal with respect to the voltage phase on the thyratron plates. Rotation speed is controlled by the magnitude of the signal.

The duration of plate current flow varies in accordance with the magnitude of the input difference signal. A low-value signal allows the tube to remain at firing potential only a short period of time. To obtain a longer firing duration, some amplifiers use a means of phase-shifting control. A separate voltage, 120° out of phase with the plate transformer voltage, is added to the difference signal, allowing the grid to rise to firing potential sooner.

Q8-16. A thyratron conducts (more, less) current than a vacuum tube.

Q8-17. What is the difference between the schematic symbols for a thyratron and a vacuum tube?

Q8-18. Firing potential of a thyratron is determined by the ____ and _____ potentials.

Q8-19. When E_1 is negative to positive (top to bottom) and the grid of V_1 is going positive, which tube will conduct? (See Fig. 8-10.)

ELECTROMECHANICAL DC SERVO AMPLIFIERS

When large amounts of power are required to move a load, electromechanical dc servo amplifiers are used. Two of these are the **Ward Leonard system** and the **amplidyne.**

The Ward Leonard System as an Amplifier

The principles of a Ward Leonard system were explained in a previous chapter. When used in a servo system, it is connected as shown in Fig. 8-11. The field winding of the dc gen-

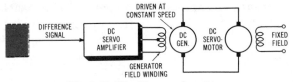

Fig. 8-11. Ward Leonard system.

erator is in the output circuit of a vacuum-tube amplifier. The output of the generator is controlled by the strength of its magnetic field. The value of current through the field will be determined by the size of the amplified difference signal. A large difference signal will cause the servomotor to attain high speed; a smaller signal, a lower speed.

When the phase of the difference signal reverses, the output polarity of the servo amplifier will also reverse, changing the direction of the magnetic field in the generator. The polarity of the generator output will reverse and the armature of the servomotor will change its direction.

An Amplidyne as an Amplifier

An **amplidyne** is a modified dc generator capable of power amplification of more than 10,000 times. A dc generator, described in an earlier chapter, uses about 100 watts of power in its field coils to generate 10 kilowatts of power. During its operation, the rotating armature develops a large reaction flux at right angles to the field. By short-circuiting the armature at this point, as shown in the diagram of Fig. 8-12, and by reducing the power applied to the field to 1 watt, the same amount of power can be developed as before.

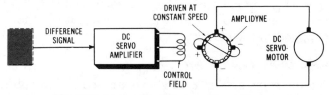

Fig. 8-12. An amplidyne in a servo system.

The amplidyne control field is split into two separate windings and is connected to the output of the servo amplifier. With no difference signal, the voltages across the two coils are equal and opposite, causing no generator output. With a difference signal fed to the servo amplifier, its output and the coil voltages become unbalanced. The coil with the higher voltage determines the direction of the magnetic field. The amplidyne amplifies the power of the field in exciting the armature of the servomotor.

Q8-20. Number the following dc servo amplifiers in descending order of power output level. The device having the largest output should be labeled No. 1.

Thyratron amplifier   Amplidyne
Transistor amplifier   Ward Leonard system

Q8-21. The Ward Leonard system consists of a(n) _ _ generator and a(n) _ _ servomotor.

Q8-22. An amplidyne has _ _ _ pairs of brushes; one pair is _ _ _ _ _ - _ _ _ _ _ _ _ _ _ _ to increase the output.

AC SERVO AMPLIFIERS

In most dc applications, the excitation voltages are applied to the servomotor armature; in ac servomotors, the voltages are applied to the stator.

A Basic Ac Servo Amplifier

The diagram in Fig. 8-13 shows an ac servo amplifier connected to an ac servomotor (a two-phase induction type).

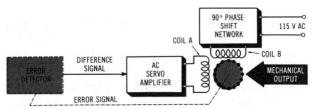

Fig. 8-13. A basic ac servo amplifier and motor.

Coil B, the reference winding, is excited from an ac line through a 90° phase-shifting network. Since the same line feeds the rest of the servo system, the difference signal, in phase or 180° out of phase with the line, will either lead or lag the voltage in coil B by 90°. The servo amplifier amplifies the difference signal before it is applied to coil A. Since the magnetic fields in the coils are 90° out of phase, the motor will rotate in a direction to correct the error signal fed back to the error detector. If the phase of the difference signal changes, the magnetic field in coil A reverses, changing the direction of motor rotation.

Types of Ac Servo Amplifiers

Ac servo amplifiers are always of the electronic type. The three most predominant types of circuits used include thyratron and multistage and single-stage vacuum-tube amplifiers.

Thyratron Amplifier—Like its dc counterpart, the ac amplifier employs thyratrons in pairs. However, only one of the tubes will conduct during any one phase of the difference signal.

Multistage Vacuum-Tube Amplifier—There are usually at least three stages in such an amplifier. The first stage adjusts the phase of the difference signal to ensure that the two induction-motor fields will be 90° out of phase. The second is a stage of phase inversion to permit the final stage to operate as a push-pull amplifier for maximum output.

Single-Stage Vacuum-Tube Amplifier — In servo systems where power output requirements are low, a cathode follower with one of the induction-motor coils in its cathode circuit will serve the purpose.

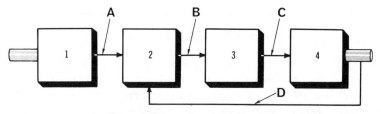

Fig. 8-14. Sketch for Problems Q8-23 through Q8-28.

Q8-23. Provide the proper titles for the numbered blocks of the servo-system diagram shown in Fig. 8-14.

Q8-24. What is the signal that appears at point A (Fig. 8-14)?

Q8-25. What determines the magnitude of signal B?

Q8-26. The name of signal D is _ _ _ _ _ _ _ _ _ _ _.

Q8-27. In most applications, the output of block 3 is a(an) _ _ or pulsating _ _ voltage.

Q8-28. Block 4 can be an ac _ _ _ _ _ _ _ _ _ motor or a dc _ _ _ _ _ - _ _ _ _ _ motor.

Q8-29. Electronic servo amplifiers employ either _ _ _ _ _ _ _ _ _ _ or _ _ _ _ _ _ _ _ _ _ _.

INPUT CONTROL FUNCTIONS

Thus far, you have learned how the output control devices (servo amplifiers and motors) perform the function of positioning a load in response to a signal from the input control section. The question that remains to be answered is how the input control devices detect and match the input and error signals to generate a difference signal.

Input Control Devices

Earlier in this chapter, you were introduced to a hypothetical ship-steering system. A simplified version of the system is shown in Fig. 8-15.

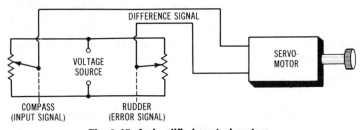

Fig. 8-15. A simplified control system.

Synchromechanisms as Input Control Devices

Although a potentiometer is used for input and error detection for many small servo systems, there are devices that accomplish these functions better. The one most widely used is a device called a **synchro**, or **synchromechanism**. A synchro is basically a transformer that has one of its windings free to rotate. Using this principle, a combination of synchros can be used to transmit positional data electrically from one location to another or used to detect and compute the difference existing between two or more shaft positions. There are five different types of synchro units.

Synchro Transmitter—This unit is sometimes called a **synchro generator**. When its **rotor** (rotating winding) is turned mechanically, the generator develops a set of voltage signals which identify the position of the rotor shaft.

Synchro Receiver—This receiver is sometimes called a **synchro motor, repeater,** or **follower**. It has the same electrical construction as the synchro transmitter and receives voltage signals to position its rotor at the same positional angle as the transmitter rotor.

Differential Synchro Transmitter—This transmitter develops and transmits the sum or difference (depending on connections) of an electrical and a mechanical input signal.

Differential Synchro Receiver—This receiver develops a mechanical (rotational) output representing the sum or difference of two electrical input signals.

Synchro Control Transformer—This transformer has both an electrical and a mechanical input. It computes, in the form of an electrical signal, the positional difference of the two inputs. The output can be used as the input to a servo amplifier.

Q8-30. **In the diagram shown in Fig. 8-15, the potentiometer arms are selecting the same amount of resistance. The servomotor (will, will not) turn.**

Q8-31. **A synchro unit acts like a(n) _ _ _ _ _ _ _ _ _ _ with a(n) _ _ _ _ _ _ _ _ winding.**

Q8-32. **Which synchro unit can be used to transmit a compass direction to a remote location?**

Q8-33. **Which synchro unit can be used as an error detector in a servo system?**

SYNCHRO FUNDAMENTALS

You have learned that a synchro unit is constructed physically as a generator (or motor) but operates electrically as a transformer. Fig. 8-16 shows the construction details of a synchro transmitter. A synchro receiver is identical to it with the exception that a receiver has a mechanical damper to prevent free-running or oscillation of its rotor. Also, the differential units and the control transformer have three rotor windings instead of one.

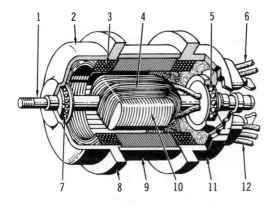

1 - ROTOR
2 - UPPER END CAP
3 - STATOR COILS
4 - ROTOR COILS
5 - BEARING
6 - ROTOR LEAD
7 - BEARING
8 - MOUNTING FLANGE
9 - SHELL
10 - ROTOR
11 - LOWER END CAP
12 - STATOR LEADS

Fig. 8-16. Synchro-transmitter construction.

Magnetic Fields in a Synchro

In the illustration given in Fig. 8-17, the schematic symbol for a synchro transmitter or receiver is shown in Fig. 8-17A. Fig. 8-17B shows bar magnets substituted for the rotor and

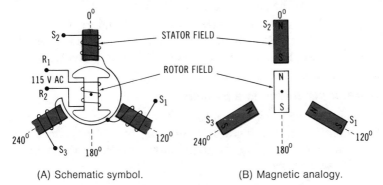

(A) Schematic symbol.   (B) Magnetic analogy.

Fig. 8-17. Representation of a synchro unit.

stator fields. Windings S_1, S_2, and S_3 are the three stator windings of a synchro. They are physically displaced 120° apart, or at equal distances around a circle. The rotor is free to rotate and is said to be positioned at 0° when the axis of the rotor is in line with the axis of winding S_2, as shown in Fig. 8-17A.

Assuming that the windings are energized, their magnetic fields can be represented by the bar magnets shown in Fig. 8-17B. Regardless of its angular position at the time the fields are energized, the rotor will turn to the position shown. The north pole of the rotor will be attracted to the south pole of S_2. The north poles of S_1 and S_3 are at equal distances to the left and right of the rotor and will pull the south pole of the rotor to the 180° position. No matter how the rotor is manually rotated, it will return to the position shown. If the polarities of all four magnets are reversed at the same time, as in the case of ac magnetic fields, the rotor will remain in the same position. If the three stator magnets are rotated together to the left or right, the rotor magnet will follow the rotational movement.

Q8-34. If magnet S_2 in Fig. 8-17B is removed, in which direction, if any, will the rotor turn?

Q8-35. If magnet S_1 is removed, in which direction, if any, will the rotor turn?

Transformer Action in a Synchro Unit

It is apparent that the rotor will align itself with the resultant field of the three stator windings. In the diagram of Fig. 8-18, the rotor winding is the primary and the stator windings are the secondary.

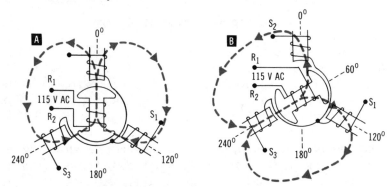

Fig. 8-18. Transformer action in a synchro transmitter.

In part A of Fig. 8-18, the rotor is positioned at zero degrees. When 115 V ac is applied to the rotor winding (the primary), a magnetic field is developed. As in any transformer, lines of flux cut the secondary windings and induce a voltage in them. The amount of voltage induced depends on the relative position of the rotor and stator windings. In part A, maximum voltage is induced in the S_2 winding because the rotor winding is exactly parallel to it. (Lines of flux are cutting the coil turns at right angles.) Windings S_1 and S_3 have a smaller induced voltage since the rotor winding and its magnetic field are at a 60° angle to their axes. The resultant induced magnetic field is shown by broken lines.

In part B of Fig. 8-18, the rotor is at the 60° position and is aligned parallel to the S_3 winding. If the stator terminals are again short-circuited to permit current flow, maximum voltage will be developed across S_3. Since the rotor is now displaced 60° to either side of S_1 and S_2, lesser but equal voltages are developed in their windings. The resultant magnetic field is shown by broken lines.

Now, imagine the rotor positioned at 30°. In this position, it is 90° from (at right angles to) the S_1 winding. The lines of flux are parallel to the S_1 coil turns and not cutting across them. No current or voltage is induced. The rotor axis is 30° from S_2 and S_3, and lines of flux are cutting across these windings at an angle that is 30° less than maximum (right angles). The voltage induced is greater than that of the 60° rotor position in part B of Fig. 8-18.

In Fig. 8-19, the stator windings of two synchro units are connected together. Because of the position of the rotor in the

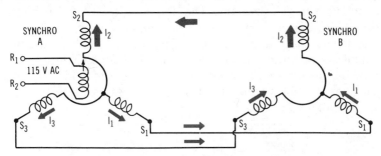

Fig. 8-19. Current flow in two connected synchro units.

left-hand synchro unit (synchro A), maximum voltage is developed across S_2 and lesser voltages across S_1 and S_3. The voltages are such that induced current I_2 is equal to the sum of I_1 and I_3 induced in S_1 and S_3, respectively. Identical currents are flowing through the respective stators of synchro B. As a result, the voltages developed across the synchro-B stator windings are equal to those across the windings of synchro A.

Q8-36. Stator windings are placed ___ degrees apart.

Q8-37. The direction of the resultant magnetic fields in the two synchros in Fig. 8-19 (will, will not) be the same.

Q8-38. Why is ac instead of dc used in a rotor winding?

A SYNCHRO TRANSMITTER-RECEIVER SYSTEM

The magnetic field produced by the primary of a transformer induces a current in the secondary. Current in the secondary winding develops a magnetic field that is opposite in polarity to the field in the primary. The diagram in Fig. 8-20 shows a similar action occurring in a synchro. Direction of the rotor field is shown by a dark arrow. The induced fields in the individual stator windings are represented by small white arrows showing a direction in opposition to the rotor field. The large white arrow shows the direction of the resultant stator magnetic field.

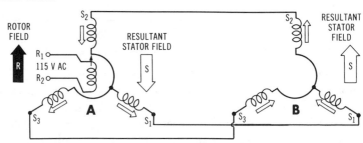

Fig. 8-20. Comparison of resultant magnetic fields.

In synchro A, the direction of the developed rotor field is up, as shown. The induced magnetic fields in the individual stators are along their axis but in a downward direction. The combination of all the stator fields produces the resultant field shown by the large white arrow. It is an induced field and is directly opposite to the rotor field.

The induced stator currents flowing in synchro B take an opposite direction to those in synchro A. Therefore, the individual stator fields produced are also in an opposite direction. And, the resultant stator field is in an opposing direction to its counterpart in synchro A. Therefore, the synchro-B stator field takes the same direction as the synchro-A rotor field.

Fig. 8-21 illustrates a synchro transmitter and receiver. Note that the two rotors are connected to the same ac source. Since

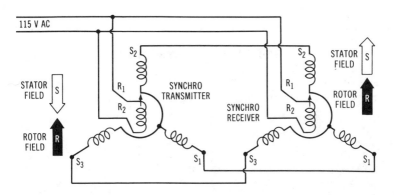

Fig. 8-21. Magnetic fields in a transmitter-receiver system.

the rotor currents are in phase, the directions of their magnetic fields will be identical. The stator field in the transmitter is opposite to the rotor field. However, the stator field of the receiver will be in the same direction as the transmitter rotor field. The rotor field of the receiver will be attracted to align itself in the stator field direction.

If the transmitter rotor is turned, the stator fields in both synchros will rotate the same amount. The receiver stator field will maintain the same direction as the transmitter rotor field. Thus, the rotor of the receiver will follow the stator field.

Q8-39. If a synchro transmitter rotor is pointing toward 90°, the transmitter stator field is pointing toward ___, the receiver stator field is pointing toward ___, and the receiver rotor field is pointing toward ___.

Q8-40. For proper operation, the _ _ _ _ _ _ of the transmitter and receiver must be connected to the same voltage source.

DIFFERENTIAL SYNCHROS

Differential synchro transmitters and **receivers** are similarly constructed and have a three-winding rotor.

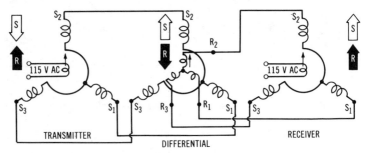

Fig. 8-22. Transmitter-differential-receiver synchro system.

Zero position for a synchro differential is where R_2 lines up with S_2. The stator windings of the transmitter develop a magnetic field in the differential stators as shown. This field induces an opposing differential rotor field which develops a receiver stator field in the opposite direction. The interaction of these fields when the transmitter and differential rotors are turned causes the receiver rotor to turn to the difference between the other two rotor positions.

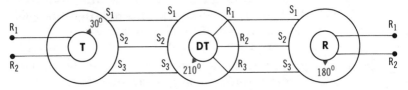

Fig. 8-23. Subtraction by a T-DT-R synchro system.

With T and DT motors mechanically turned to 30° and 210°, the interacting magnetic fields will cause rotor R to electrically turn to the difference position, 180°. Differential transmitters will add if connected as shown in Fig. 8-24. Here, S_1 and S_3 of T are crossed to S_3 and S_1 of DT. R_1 and R_3 of DT are crossed to S_3 and S_1 of R. The angular positions of 75° on T and 175° on DT are added to produce an R rotor position of 250°.

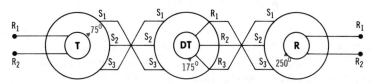

Fig. 8-24. Addition by a T-DT-R synchro system.

When connected to two synchro transmitters, the DR rotor is the sum or difference of the rotor positions of the two generators. The Fig. 8-25 synchros are connected for subtraction.

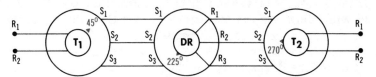

Fig. 8-25. Subtraction by a T-DR-T synchro system.

Addition, with a differential receiver recording a sum of two angles, is accomplished by the following connections.

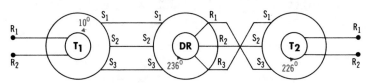

Fig. 8-26. Addition by a T-DR-T synchro system.

Q8-41. In a T-DR-T system, the inputs to the DR are (electrical, mechanical) and its output is

_ _ _ _ _ _ _ _ _ _ .

Q8-42. In a T-DT-R system, one DT input is electrical and the other mechanical; the output is

_ _ _ _ _ _ _ _ _ _ .

THE SYNCHRO CONTROL TRANSFORMER

A control transformer (CT) has three stator windings. Its rotor is designed to generate a voltage, the amplitude of which represents the difference between two angular positions and the phase of which shows the direction of difference.

The rotor of a control transformer is at zero position when it is at right angles to S_2 as shown in the diagram of Fig. 8-27. The induced stator field in the CT will be in the same direction as the rotor field of T. In the positions shown, no voltage will be produced between R_1 and R_2 of the CT.

If the T rotor is moved in a clockwise direction, the CT **stator** field will follow. As the stator field of the CT rotates, its flux lines cut the rotor winding. Rotation in a clockwise direction will produce a voltage in the rotor in phase with the alternating current on the T rotor. When the CT stator field is 90°, the voltage between R_1 and R_2 is maximum. A 360° rotation of the stator field will produce an ac sine wave.

If the T rotor is turned in a counterclockwise direction, the flux lines of the CT stator field will cut its rotor winding in the opposite direction. This will produce a voltage across the CT rotor 180° out of phase with the T rotor voltage.

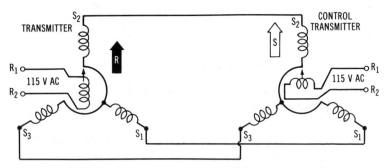

Fig. 8-27. A synchro transmitter-control transformer system.

The rotor of the control transformer will not turn as the result of the moving stator field. It can be rotated by manually turning the rotor shaft. The diagram of Fig. 8-28 demonstrates what occurs when the rotor is turned.

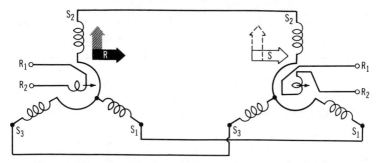

Fig. 8-28. Rotating CT rotor to reduce voltage output to zero.

Assume that the T rotor has been turned clockwise from $0°$ to $90°$. The CT stator field will make the same rotation. If the CT rotor remains at $0°$, maximum voltage in phase with the ac line voltage can be measured across R_1 and R_2. If the CT rotor is now turned to $90°$, the voltage reading will reduce to zero. In this position, the rotor winding is at right angles to the stator field.

It is evident that this type of synchro system can be used to develop the difference voltage for a servo system. The synchro transmitter rotor can reflect the desired angular position and the CT rotor the present position of a load. If there is a difference in positions, the magnitude of the voltage on the CT rotor will be proportional to the amount of difference, and the phase of the voltage will reveal the direction of difference.

If more than two position readings are required in a servo system, a differential transmitter can be substituted in place of the synchro transmitter. The output of the DT to the CT will be the sum or difference (depending on the stator and rotor connections) of two position conditions.

Q8-43. When the stator field of a CT rotates in a(n) _ _ _ _ _ _ _ _ _ direction, the developed voltage is in phase with transmitter rotor voltage.

Q8-44. Maximum voltage is developed if the stator field and rotor winding are (at right angles, parallel).

THE COMPLETE SERVO SYSTEM

The diagram in Fig. 8-29 illustrates a complete system. The

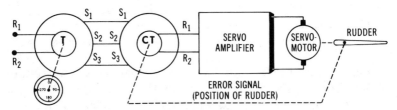

Fig. 8-29. Synchro system used as input control.

rotor of the synchro transmitter is physically connected to a compass. As the ship swings right or left of the established course (set at rotor zero position), the stator field of the control transformer will follow the transmitter rotor. This will generate a voltage across the rotor of the control transformer. The amplitude of the voltage will be proportional to the difference between compass and rudder positions. The phase will show whether the compass has rotated clockwise or counterclockwise.

The voltage output of the CT rotor is fed to the servo amplifier as the difference signal. This unit amplifies the signal to the level of power required to operate the servo motor. The amplitude of the signal will determine how fast the motor will turn the rudder. The phase of the signal will control the direction in which it will be turned.

When the rudder angle has matched the off-course angle of the compass, the motor will stop. Since the rudder is connected to the rotor of the control transformer, an error signal is returned to this synchro unit. As the rudder is turned to its desired position, the CT rotor is approaching right angle (zero voltage) alignment with the stator field. This causes the difference signal to decrease in amplitude until zero voltage is reached when compass and rudder angles are the same.

The rudder angle causes the ship to swing back to its course. A new difference signal is generated in the opposite phase. The motor turns the rudder in the opposite direction and toward amidships again. When the ship is finally on course, transmitter and control transformer rotors are in alignment and the difference signal is zero.

SERVO SYSTEM APPLICATIONS

Commercial and military aircraft use servo systems to operate the control surfaces of airplanes and as an automatic pilot to keep the plane on a desired course and altitude. Servos are also used in analog computers to detect and compute rates of change of quantities. With an amplidyne, industry uses servo systems to move and position heavy loads.

Military applications include the positioning of radar and heavy directional radio antennas, the pointing of guns and missile launchers, the automatic control of missile steering mechanisms, and other uses requiring precise positioning of a load or rapid data computation.

Q8-45. In Fig. 8-29, the _ _ _ _ _ _ _ _ _ _ _ _ _ _ _ _ _ is an input detector, and the _ _ _ _ _ _ _ _ _ _ _ _ _ _ _ _ is an error detector.

Q8-46. Since the servo amplifier output is being applied to the armature of the servo motor, the motor is (dc, ac).

Q8-47. If the motor were (dc, ac), the output would be applied to one of two fields in the motor.

Q8-48. The following synchro units have three windings on their rotors: (A) _, (B) _.

Q8-49. The following synchro units have a single winding on their rotors: (A) _ _ _ _ _ _ _ _ _ _ _ _, (B) _ _ _ _ _ _ _ _, (C) _ _ _ _ _ _ _ _ _ _ _ _ _ _ _ _.

WHAT YOU HAVE LEARNED

1. A servo system is a combination of devices that permit automatic control and positioning of a load. Each system (large or small) includes two functions—input control and output control. The former detects a deviation between the desired and the existing position of a load and sends a correcting signal to the output control function which realigns the load.

2. The input control function contains an input detector and an error detector. The output control function has a servo amplifier and a servomotor. Two of these operations can be obtained within a single device, or several mechanisms may be required to just perform a single operation.

3. An open servo system has some of its operations performed by manual or semiautomatic means. A closed system is completely automatic.

4. An ac servomotor (normally an induction type) is used in a control system when low power and low speed are permissible. High torque and/or a wide speed range will call for either a shunt motor or permanent magnet dc motor.

5. The purpose of a servo amplifier is to convert a difference signal into sufficient power and polarity to turn a servomotor in the correct direction at the desired speed.
6. Electronic servo amplifiers (vacuum tube or thyratron) can be designed to drive either ac or dc servomotors.
7. When large amounts of power are required to move a load in a servo system, electromechanical dc servo amplifiers are used. Examples are the Ward Leonard dc generator-motor and the amplidyne. Because it has a short-circuited generator armature, an amplidyne has a power amplification factor of up to 10,000, or greater.
8. The most widely used device to perform the input control function of input- and error-signal detection is the synchro. A synchro unit is basically a transformer with one of its windings free to rotate. Synchro units may also be used to transmit data between several remote locations.
9. A synchro transmitter or receiver contains three stator windings (placed 120° apart) and a rotor with a single winding. If alternating current is applied to the rotor, the magnetic field induces current and a magnetic field in the stator windings. The rotor and stator fields lie along the same axis but take opposite directions. If the stator terminals of two synchro units are connected, the induced stator field of the second unit will take the same direction as the rotor field of the first unit.
10. A data-transmission system can be formed by connecting the stator windings of a synchro transmitter to those of a synchro receiver. Since the receiver stator field follows the rotation of the transmitter rotor field and since the receiver rotor field will line up with its stator field, the receiver will precisely duplicate the angular position of the transmitter.
11. Differential synchro transmitters and receivers have three stator windings and three rotor windings. A differential transmitter receives an electrical and a mechanical input and delivers an electrical output. A differential receiver accepts two electrical inputs and delivers a mechanical output. When connected to other synchro units, a differential will develop the sum or difference of two positions, depending on the connections.

12. A control transformer has three stator windings and a single-coil rotor that will develop a voltage of an amplitude proportional to the difference of its rotor and stator fields, and of a phase representative of the direction of difference. When a synchro transmitter is connected to a control transformer, the pair can perform the input control function of a servo system.

Index